Mathematical Tools for Real-World Applications

T0091945

Mathematical Tools for Real-World Applications

A Gentle Introduction for Students and Practitioners

Alexandr Draganov

The MIT Press
Cambridge, Massachusetts
London, England

©2022 Massachusetts Institute of Technology

All rights reserved. No part of this book may be reproduced in any form by any electronic or mechanical means (including photocopying, recording, or information storage and retrieval) without permission in writing from the publisher.

The MIT Press would like to thank the anonymous peer reviewers who provided comments on drafts of this book. The generous work of academic experts is essential for establishing the authority and quality of our publications. We acknowledge with gratitude the contributions of these otherwise uncredited readers.

This book was set in Times Roman by the author. Printed and bound in the United States of America.

Library of Congress Cataloging-in-Publication Data

Names: Draganov, Alexandr, author.
Title: Mathematical tools for real-world applications : a gentle introduction for students and practitioners / Alexandr Draganov.
Description: Cambridge, Massachusetts : The MIT Press, [2022] | Includes bibliographical references and index.
Identifiers: LCCN 2021046200 | ISBN 9780262543965 (paperback)
Subjects: LCSH: Mathematics. | Problem solving—Mathematical models.
Classification: LCC QA39.3 .D73 2022 | DDC 510—dc23/eng/20211119
LC record available at https://lccn.loc.gov/2021046200

10 9 8 7 6 5 4 3 2 1

To Luda

The reader who follows my drift with sufficient attention will easily see that nothing is less in my mind than ordinary Mathematics, and that I am expounding quite another science, of which these illustrations are rather the outer husk than the constituents. Such a science should contain the primary rudiments of human reason, and its province ought to extend to the eliciting of true results in every subject.
—René Descartes, *Rules for the Direction of the Mind*

Indeed, I have found that it is usually in unimportant matters that there is a field for the observation, and for the quick analysis of cause and effect which gives the charm to an investigation.
—Arthur Conan Doyle, *The Adventures of Sherlock Holmes*

Contents

List of Figures

List of Tables

Preface

Real scientific or engineering problems are often different from textbook ones and take a different set of tools, for two reasons.

First, real problems are usually messier and take longer to solve. A good sprinter is not necessarily a good marathoner; similarly, a long solution requires a different set of skills than a short one. You must plan out an approach in advance, break it up into manageable segments, and devise ways to check your work along the way. In math, this starts from understanding your problem well.

Second, in contrast to many textbook problems, an engineer or a scientist rarely seeks a solution in the form of a *number*. This is because a numerical result is usually not a goal in itself; rather, one may be designing a better device, or modeling a process, or interpreting available data. Achieving these goals requires exploration of different options and studying trade-offs. This means that a solution is treated as a box with knobs to turn and levers to pull. One tries different things and sees what happens. A single numerical output simply does not cover the entire range of possibilities.

This book is written with these two considerations in mind. It will teach you mathematical tools to deal with real scientific or engineering problems. Rather than applying standard recipes for solving a problem, we will concentrate on making sense of a solution by analyzing it in different ways. (To be fair, not all of this book is about analyses: we do study one method to solve some tough problems.)

In addition, you will learn to better see the connections between a scientific or engineering problem in hand and the equations that describe it. A mathematical representation for a real-life problem should be a working mathematical model of that problem. Just as a model toy car emulates the real car's running and steering, a mathematical model should bear some essential features of the system it depicts. Seeing the mapping between the real system and its math model is an essential skill for any scientist or engineer.

Upon reading this book, you will:

- Learn to *understand a problem* better and to see the connection between its formulation and the equations. This will also tell you what to expect from the final solution and will help you select a path to it.
- Discover a new way to *solve* tough problems that cannot be cracked by standard tools.
- Learn how to *test and validate* your solutions. In the "real world" the cost of an error in a solution can be very high, and confidence in your result is important.
- *Determine* the boundaries of validity for your solution: establish where it works and where it doesn't. For any problem, you need to make sure that your solution applies to the situation in hand. In addition, it is often tempting and rewarding to look beyond the boundaries – that may be a way to get new ideas.

The material is presented using several dozen sample mathematical problems. These problems are not the essence of this book; they just serve as fodder for discovery. This book presents a hybrid approach: we will use some fairly standard problems that introduce and illustrate new mathematical tools. The problems may look familiar, but tools for dealing with them may be new to you. We will use only algebra and trigonometry; knowledge of calculus may be helpful but is not required for better understanding some sections.

The book has eight chapters. We start with an introduction to units and dimensionality. It is surprising how often people fail to check units in their equations, even though this is a simple and powerful way to test the validity of a solution. In some cases, unit analysis can even produce a solution.

Next, we'll explore how to use special or extreme values of parameters to gain insight into a problem. This technique serves two purposes: it helps us understand our problem by defining its domain, and it is a great way to point to errors in the solution. If we think of a problem as a country, limiting cases are like that country's borders: we can peek outside and see what's out there.

The next two chapters discuss symmetry and scaling, exploring some simple but surprisingly useful techniques. For example, we may want to explore what happens if the values of two parameters are swapped, or if a parameter is scaled by a constant factor. For many problems, the results may be intuitively obvious from the start: we just *know* what to expect. It is always useful to find a manifestation of symmetry or scaling in the equations. This helps us test the final result and better understand it.

The fifth chapter shows how to roughly estimate a solution. Most real-life problems are affected by countless factors. A comprehensive treatment could be very complex and may not even be feasible. A common way to deal with these issues is to neglect some weak factors that don't affect the result much. For example, an engineer computing a satellite orbit must account for Earth's gravity. But should she account for the gravity from the Sun and the Moon? Should she account for the nonuniform distribution of mass in Earth? Before plunging in complex computations, she may want to estimate the rough magnitude of each phenomenon and draw a line for dealing with only some of them.

The sixth chapter introduces successive approximations. This is a very powerful method to crack some problems that seem unsolvable otherwise. Technically, successive approximations do not "solve" the problem, in the narrow sense of the word, but they often produce a solution that is "good enough" when there is no better alternative anyway. We explore this method using simple and intuitive algebra. A knowledge of calculus is not necessary but would be helpful to fully appreciate the intricacies of this method.

The seventh and eighth chapters present two no-kidding, real-life problems: computing the probability of a rare, catastrophic event, and tracking a satellite with a GPS receiver. We will examine only the results and do not go through the derivations because they are quite complex. Yet we analyze, dissect, and gain insight into the results by using all the tools that we have explored in the previous six chapters. Chapters 7 and 8 tie everything together and show how analytical tools work on problems that were not concocted for a textbook. Mathematical treatment in these chapters progresses from relatively straightforward (chapter 7) to more challenging (chapter 8). The equations in chapter 8 are more cumbersome than those in the previous chapters, but this complexity is what one encounters in many real-life applications. It is precisely when dealing with complex equations and long derivations that quick analyses and validity checks become most useful.

Finally, the appendix presents solutions for most of the problems that are scattered throughout the book. Since you'll often apply multiple techniques from this book to any given problem, many problems in the appendix are used to illustrate more than one concept. This serves two purposes: (1) you will see how different tools work on the same problem, and (2) you have to learn only once about inconsequential specifics of each problem, and then you can better concentrate on the subject of each chapter where that problem is mentioned.

Most of the techniques presented in this book have been in the quiver of professionals for decades, if not centuries, but they are rarely taught in school. Often, scientists and engineers learn them on the job or from their colleagues. The goal of this book is to give you a primer on six powerful analytical tools that you will be able to use over and over, whether you are designing a driverless car, planning the deployment of a telecommunications network, or parsing data from a clinical trial.

This book derives from what I learned over the years from many great teachers, including Ms. Vera Rosenberg and Drs. Yuri Gorgo, Nickolay Kotsarenko, Umran Inan, and Tim Bell. I received many helpful comments on the manuscript from Dr. Oleg Lavrentovich, Ms. Maria Shmilevich, and Messrs. Greg Harrison, Andrii Denysenko, and Daniel Heitz. I am grateful to Dr. Boris Veytsman for his expert help with all things LaTeX. Writing this book would not be possible without the patient support and help from my wife, Luda, our children, Anna and Andrew, and our son-in-law, Daniel Palmer. The work of my father, Boris Draganov, and the books published by him were my inspiration for this project.

How to Read This Book

Before you dive into this material, let me offer you a couple of words about how to tackle this text.

Each of the first six chapters contains a brief explanation of a mathematical tool, followed by demonstrations of how this new tool applies to various problems. These problems are the meat of this book. Most problems are "recycled": the same problem is referenced in several chapters to illustrate different topics. Solutions for most of these problems are provided in the appendix, and the text in the first six chapters references these solutions.

Some of the tools presented in this book are helpful in finding mathematical errors. In practice, almost anyone can make a mistake, especially when dealing with a long and complex derivation. In addition to inspecting your work step by step, it is always helpful to have tools that can quickly flag potential issues. To illustrate this, we occasionally explore a flawed solution and then apply a particular tool to detect the flaw. In these cases, at the start of erroneous equations I have put a small Pinocchio icon facing right, and at the end the same icon facing left (as in 🐢 $2+2 = 5$ 🦞). Sometimes, an error is planted in only a part of a derivation. Since our purpose is to learn how to find hidden errors, I enclose the entire derivation between two Pinocchios if it contains an error or errors anywhere, even if some of its equations are correct.

Each of the first six chapters is followed by a set of exercises. Some of the exercises are multiple choice, where you shall discriminate between correct and wrong formulas. Other problems require producing a mathematical result, a free-form explanation, or both. Some problems are marked with an asterisk (*); these are designed to challenge you more.

The last two chapters have no exercises because they do not present new material. As mentioned in the preface, chapters 7 and 8 apply the mathematical techniques presented in chapters 1–6 to two tough real-life problems. These chapters show how these techniques work together and demonstrate their utility.

The appendix presents solutions to many of the problems analyzed in different ways in chapters 1–6. You may be able to learn great mathematical techniques without ever peeking into the appendix, but going through it would still be useful: it will help you see the intricate connections between the different ideas you will have learned from this book. Some of the problems solved in the appendix may be familiar from standard mathematical curriculum. In this book they serve as fodder for the material in chapters 1–6 and for the exercises. All problems in the appendix and some in the exercises for chapters 1–6 are used more than once; such problems are referenced in a separate index of problems.

A word of caution: since problem examples are drawn from many disciplines, it is impractical to ensure a full consistency of notations throughout the book. Notations are consistent within each section, but not across different sections. For example, μ may denote the offset in the Pareto distribution in chapter 7 and, separately, a small parameter for Earth's deviation from the spherical shape in section 6.7.

The mathematical tools presented in this book are universal: their applications span wide ranges in subject matter and in complexity. The examples in the book are deliberately limited to simpler problems. Most of the material requires just knowledge of algebra and trigonometry, in the hope that this helps you concentrate on the subject and spend less effort going through advanced mathematical calculations. Some knowledge of calculus will be helpful (though not required) for chapters 6–8.

1 Units

Many real-life quantities are associated with some units of measurement. For example, distance may be measured in meters, and time in seconds. A unit of measurement is defined by a convention or by law as a standard for measuring quantities of the same kind. A failure to standardize units and to track them in all calculations may have disastrous consequences. In 1999, NASA lost a $125-million Mars Climate Orbiter because engineers at the Jet Propulsion Laboratory and at Lockheed Martin Aeronautics failed to reconcile measurement units for the thrust force of the spacecraft. This chapter explores how to use units to check results and, in some cases, even to obtain one. Leveraging units to investigate a problem is often called dimensional analysis.

Units may be base or derived. For physical quantities, there are seven base units, which are listed in table 1.1. Of course, not every problem is based on physical quantities. For example, in a financial calculation certain variables may be measured in dollars.

Key Point

Units must be consistent in all equations.

Table 1.1
Base units for physical quantities

Quantity	Name	Symbol
Length	meter	m
Mass	kilogram	kg
Time	second	s
Electric current	ampere	A
Temperature	kelvin	K
Amount of substance	mole	mol
Luminous intensity	candela	cd

Derived units are produced from the base units as a result of multiplication, division, and/or exponentiation. Take a car that goes 60 meters in 2 seconds; its speed is computed as $60/2 = 30$ m/s; hence the units for the speed are m/s. Sometimes, derived units have their own name. For example, power is measured in watts (W), where a watt is $kg \cdot m^2 \cdot s^{-3}$. Keep in mind that the same quantity may be measured by more than one type of unit: power can also be measured in horsepower (1 hp $\approx$ 745.7 W).

Note that in the car speed example we expressed the value in the units of meters *per* second. Though it is not a formal rule, units for a quantity that is computed per unit of another quantity are typically expressed as fractions. This applies to various density units, such as the density of a substance, which is the mass per unit volume (such as kg/m^3).

Units must be consistent in every equation. The rules for unit consistency are as follows:

1. Additive terms must have the same units. For example, a sum of distances d_1 and d_2 is $d_1 + d_2$. Both can be measured in meters or in miles; we may not measure d_1 in meters and d_2 in miles.

2. Units of a sum or a difference of two quantities are the same as the units for each of the terms.

3. Both sides of any equation must have the same units. (Note that, in dimensional analysis, the common saying "time is money" is wrong!)

4. For multiplications and divisions units obey regular algebraic rules. For example, if a hiker walked distance d miles (mi) in t hours (hr), then her average speed was $V = d/t$, which is measured in miles per hour (mi/hr).

5. The last rule means that a power of any quantity has the corresponding units raised to the same power. For example, if a side of a square is r meters (m), then its area is r^2 and is measured in m^2. This rule applies to noninteger exponents as well, including radicals.

6. Nonzero numerical constants are dimensionless. For example, an area of a circle with radius r is $A = \pi r^2$. For the area to be measured in m^2, the constant π must be dimensionless.

7. Arguments and values of all transcendental functions are dimensionless.[1] This applies to trigonometric functions, exponents, logarithmic functions, and many others, such as Bessel functions, the gamma function, or the error function. Note that while the base a of an exponent a^b can be measured in any unit, the exponent proper b must be dimensionless.

To illustrate the last rule, consider trigonometric functions. The value of a sine is defined as the ratio of the opposite side of a right triangle to the hypotenuse. If the sides of a

---

1. Transcendental functions are those that cannot be expressed through operations of addition, subtraction, multiplication, division, raising to a power, or computing a radical.

triangle are measured in meters, then their ratio is dimensionless, as is the value of sine. The argument of a sine is measured in radians. The radian is defined using the ratio of the length of a circular arc to the radius of the circle. Again, such a ratio (and the measure of the angle) is dimensionless. Thus, both the argument and the value of the sine function are dimensionless.

To see how these rules apply in practice, consider a formula that uses several physical quantities $x_1, x_2, \ldots, x_n$ measured in their respective units. A function of these quantities may have combinations of these quantities in a product $x_1^\alpha x_2^\beta \ldots x_n^\omega$, where $\alpha, \beta, \ldots, \omega$ are numerical constants. Combinations of this type may be added to or subtracted from each other as long as their units match. They can be used as arguments of transcendental functions as long as they are dimensionless. In these cases, the values of these transcendental functions are dimensionless as well.

The rules on unit consistency apply to every equation and every mathematical expression of a solution, from the start to the final result. For any real-life problem, checking units must become a part of a routine: it is one of the simplest ways to find certain types of errors in your solution. If the units do not match, there definitely is an error. The converse is not necessarily true: units can be consistent and the solution can still be erroneous.[2]

In many cases, it is possible to find a dimensionless combination of several quantities. Such combinations may point to important hidden features in the problem. When we study linear and quadratic scaling in chapter 4, we will learn about the Reynolds number, which illustrates this concept.

Dimensionless variables present one more important benefit: they can be characterized as "small" or "large." For a variable that is measured in any unit, such a characterization does not make sense: a numerical value for a distance can appear to be small when measured in kilometers but would seem large if measured in millimeters. For a dimensionless quantity, such as a ratio of two distances, this problem goes away. This consideration is key for making order-of-magnitude estimates, which we will study in chapter 5, and for using the method of successive approximations in chapter 6.

1.1 Using Dimensional Analysis to Solve Problems

Occasionally you may encounter a situation where only one combination of variables yields the desired units for the solution of your problem. In this case you can seek the solution with confidence as a dimensionless numerical multiplier times that unique combination of variables. Below is an example of this situation.

---

2. An expression ✍ $A = \pi r^3$ ✍ for the area of a circle is wrong: the units do not match. However, a unit check would not point to an error in the formula ✍ $A = r^2$ ✍ for the area of a circle. Though that last formula is missing a π multiplier, the units still match.

Suppose that you are tasked with computing the air drag force for a new model of a car. After some initial research, you discover that two mechanisms contribute to the air drag force: an internal friction in the air and the force that pushes the air around the car as the car moves. These two mechanisms are largely independent and can be considered separately. Here we concentrate on the second one, which dominates the air drag force at large velocities.

We can hypothesize that the result must depend on three variables:

1. The speed of the car V. Indeed, a faster-moving car must overcome a stronger air drag. Speed is measured in m/s.

2. The density of the air ρ. If the air is rarefied, the force should be smaller. A space vehicle does not slow down much over time exactly for this reason. Air density is measured in kg/m^3.

3. The size of the vehicle D. We know that it is easier to push a small object through the air than a large one. Vehicle size is measured in meters (m).

We are looking to compute the air drag force, which is measured in $kg \cdot m/s^2$. There is only one combination of the three relevant variables that yields correct units. The air drag force must be proportional to this unique combination of variables:

$$F \propto \rho D^2 V^2, \tag{1.1}$$

where $\propto$ denotes proportionality. The exact formula may be different from the right-hand side of the last equation by a constant dimensionless multiplier. For historical reasons, the canonical formula uses the cross-section area A of the object, which in turn is proportional to D^2. It also defines the proportionality coefficient as $\frac{1}{2}C_D$, where C_D is a dimensionless parameter called the drag coefficient. The final formula is given as

$$F = \frac{1}{2}C_D \rho A V^2. \tag{1.2}$$

The value of the drag coefficient C_D depends on the shape of an object. The shapes of cars and airplanes are designed to make the drag coefficient smaller. For a sail on a boat, the goal may be the opposite.

In this example, we see how a simple dimensionality argument produces a solution for a tough problem. An exact treatment requires solving the so-called Navier-Stokes equation, which is a challenge even when using modern numerical methods and powerful computers.

Key Point

If only one combination of relevant variables yields the correct units, then this combination times a dimensionless multiplier is a solution for the problem in hand.

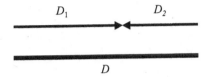

Figure 1.1
Two hikers on a trail: units

To summarize, for many applications, units give us a powerful way to check the solution at every step and occasionally even a way to predict its final form. Now we proceed to several examples, which will further illustrate applying dimensional analysis.

1.2 The Two Hikers Problem

Consider two hikers who walk toward each other (figure 1.1). One hiker maintains speed V_1, and another maintains speed V_2. The total length of the trail is D. What distances will they cover before meeting each other? Section A.1 provides a solution to this problem: hikers 1 and 2 will travel the following distances:

$$
\begin{aligned}
D_1 &= \frac{DV_1}{V_1 + V_2}, \\
D_2 &= \frac{DV_2}{V_1 + V_2}.
\end{aligned}
\tag{1.3}
$$

Let's assume that distances are measured in meters (m) and time intervals in seconds (s). Then speeds are measured in meters per second, or m/s. A check of solution (1.3) shows that units on the left-hand side and the right-hand side match.

1.3 The Circle and Line Problem

Here we find the horizontal coordinates of the intersections of a circle and a line (figure 1.2) that are given by the following equations:

$$
\begin{aligned}
x^2 + y^2 &= R^2, \\
y &= px + q.
\end{aligned}
\tag{1.4}
$$

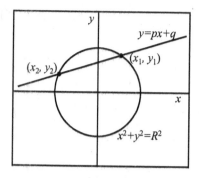

Figure 1.2
A circle and a straight line: units

Section A.3 shows that coordinates $x_{1,2}$ of the intersections (if any) are given by

$$x_{1,2} = \frac{-pq \pm \sqrt{(1 + p^2)R^2 - q^2}}{1 + p^2},\tag{1.5}$$

where subscripts $1, 2$ correspond to the $\pm$ signs in the right-hand side.

In this section, we look at the units in equations (1.4) and in their solution (1.5). Suppose the radius of the circle is measured in meters (m). In the original equations (1.4) the units must be consistent. This means that x and y must be measured in meters as well (see the first equation). If we look at the second equation, we can see that q must be measured in meters, and p must be dimensionless.

The units for all three parameters and for the unknowns will ripple through the entire solution of this problem. The same units must be consistent in the expression for the roots (1.5). Indeed, the denominator is dimensionless (note the sum of 1 and p^2). The numerator has two major terms, pq and the square root term. The first term is measured in meters (p is dimensionless and q is in meters). The expression under the square root is measured in m^2 because $1 + p^2$ is dimensionless, and both R^2 and q^2 are in m^2. Thus, the value of the square root is in meters, and the entire expression for $x_{1,2}$ is in meters.

There is another subtlety in this problem. Note that only two (R and q) of the three input parameters are measured in meters. What if we measure them in centimeters instead? The geometry of the problem would not change, but all coordinate values will be scaled by a factor of 100. As all coordinates and distances are scaled in the same way, the solution remains valid.[3] One way to achieve this effect is to express the solution through *ratios* of

---

3. We explore the concept of scaling in depth in chapter 4.

lengths; such ratios would be dimensionless and would not change whether the lengths are measured in meters or centimeters.

Let us check what happens with equation (1.5) if we express it through ratios of the quantities that are measured in meters. We divide this equation by the radius of the circle R:

$$\frac{x_{1,2}}{R} = \frac{-p\frac{q}{R} \pm \sqrt{(1 + p^2) - \left(\frac{q}{R}\right)^2}}{1 + p^2}. \tag{1.6}$$

Key Point

Try to express your solution through dimensionless combinations of variables. Often this reduces the complexity of the problem.

The benefit of this last solution is that we effectively have two parameters (q/R and p) instead of three in the right-hand side. If we want to see what happens with the solution at different values of $p, q,$ and R, we no longer have to vary all three parameters. We need to vary only p and q/R, and this would still cover the entire space of possibilities. This is very typical for many problems, and it is common to express a solution through dimensionless variables.

1.4 Satellite Coverage

Consider a satellite that flies at altitude H above Earth and transmits a signal downward (figure 1.3). The antenna beam has a conical shape, and the angle from the center to the

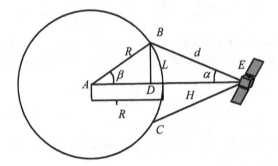

Figure 1.3
Designing satellite coverage: units

edge of the beam is α. Earth's radius is R. We need to compute the length L of the arc on Earth's surface from the center of the beam spot to its edge. This problem is solved in section A.24 to yield

$$L = R\left(\sin^{-1}\left(\frac{R+H}{R}\sin\alpha\right) - \alpha\right). \tag{1.7}$$

Are the units in this solution correct?

Earth's radius R and the orbit altitude H are measured in meters. Therefore, the expression $(R+H)/R$ in equation (1.7) is dimensionless. From the rules for unit consistency, we know that both α and $\sin\alpha$ are also dimensionless. Therefore, the argument of $\sin^{-1}$ in equation (1.7) is dimensionless, as is its value. We conclude that equation (1.7) produces the result in meters, as it should. Units in this solution are therefore consistent.

This supports the validity of equation (1.7). Of course, our unit check is not definitive, and not every error can be caught by dimensional analysis.

1.5   The Cubic Formula

Unit checks are particularly useful if equations are cumbersome and have a lot of parameters. Here we consider a general formula for solving a cubic equation for t (here we use t for the unknown, because in this section we will solve the cubic equation for time):

$$at^3 + bt^2 + ct + d = 0. \tag{1.8}$$

The solution of this equation is as follows (see section A.29). We denote:

$$\Delta_0 = b^2 - 3ac,$$
$$\Delta_1 = 2b^3 - 9abc + 27a^2d,$$
$$C = \sqrt[3]{\frac{\Delta_1 \pm \sqrt{\Delta_1^2 - 4\Delta_0^3}}{2}}, \tag{1.9}$$

where the $\pm$ sign in this solution can be chosen arbitrarily, unless $C = 0$ for one of the signs, in which case we must choose the sign that yields a nonzero value of C. Then one of the roots is given by

$$t_1 = -\frac{1}{3a}\left(b + C + \frac{\Delta_0}{C}\right). \tag{1.10}$$

This formula is rarely used in practice because of its complexity, but this very complexity also makes a unit check particularly powerful.

We consider an application where a car decelerates from some initial velocity for a red light. Most often a driver does not apply the brakes with the same force; instead, she gradually increases the pressure on the brake pedal. In our simplified model, this increase is linear in time.

We want to solve for the time that is required to come to a full stop from the moment when the stopping point is d meters away. This problem is modeled by equation (1.8), where the unknown time to the full stop t is measured in seconds.[4]

Units for the coefficients are listed in table 1.2. All additive terms in equation (1.8) are measured in meters.

Table 1.2
Units for the coefficients in the car deceleration formula

Coefficient	Physical quantity related to this coefficient	Units
d	Distance	m
c	Velocity	$\dfrac{m}{s}$
b	Acceleration	$\dfrac{m}{s^2}$
a	Jerk	$\dfrac{m}{s^3}$

Let us now see if the cubic formulas (1.9) and (1.10) have consistent units. We consider various terms and expressions in these equations, starting from the first equation in (1.9) and proceeding to equation (1.10). For each term or expression, we refer to table 1.2 to determine its units. We start from simpler terms and build our analysis up from there.

Table 1.3, which summarizes the results, shows that the units of all expressions are in exact agreement, which makes us more comfortable about this solution. Needless to say, a derivation of the cubic formula is quite cumbersome, and it is easy to make a mistake there. While today we can rely on textbooks for the correct solution, the original derivation was a challenge. In such situations, unit checks help track the validity of the solution at every step along the way. It is not a definitive test, because an erroneous equation may pass a unit check, yet it should be a tool in your arsenal. For most applications, checking units along the way while solving a problem should be a routine procedure.

---

4. In general, a cubic equation with real coefficients may have one or three real roots. For this application, we assume that the root given by equation (1.10) happens to be the one we need. If it turns out that for particular values of parameters a, b, c, and d we must deal with another root, the analysis would be similar.

Table 1.3
Units for various terms in the cubic formula

Terms	Units	Terms	Units
b^2	$\dfrac{m^2}{s^4}$	$4\Delta_0^3$	$\dfrac{m^6}{s^{12}}$
$3ac$	$\dfrac{m^2}{s^4}$	$\sqrt{\Delta_1^2 - 4\Delta_0^3}$	$\dfrac{m^3}{s^6}$
$\Delta_0 = b^2 - 3ac$	$\dfrac{m^2}{s^4}$	$\Delta_1 \pm \sqrt{\Delta_1^2 - 4\Delta_0^3}$	$\dfrac{m^3}{s^6}$
$2b^3$	$\dfrac{m^3}{s^6}$	$C = \sqrt[3]{\dfrac{\Delta_1 \pm \sqrt{\Delta_1^2 - 4\Delta_0^3}}{2}}$	$\dfrac{m}{s^2}$
$9abc$	$\dfrac{m^3}{s^6}$	$\dfrac{\Delta_0}{C}$	$\dfrac{m}{s^2}$
$27a^2d$	$\dfrac{m^3}{s^6}$	$b + C + \dfrac{\Delta_0}{C}$	$\dfrac{m}{s^2}$
$\Delta_1 = 2b^3 - 9abc + 27a^2d$	$\dfrac{m^3}{s^6}$	$t_1 = -\dfrac{1}{3a}\left(b + C + \dfrac{\Delta_0}{C}\right)$	s
Δ_1^2	$\dfrac{m^6}{s^{12}}$		

1.6 Summary

For any real-life problem that deals with measurable quantities, an analysis of units should be a matter of course. It serves two purposes:

1. We can check the solution. This check is not 100 percent definitive because an erroneous solution may occasionally pass it, but if a solution does not pass a unit check, there must be an error.
2. Often we can combine parameters to produce a dimensionless variable. If found, such a combination fully encapsulates the effects of constituting parameters. As a result, we do not have to perform analyses for individual parameters but instead can deal with a smaller number of dimensionless combinations. This tactic also illuminates how different parameters play together.

Occasionally, a unit analysis may yield a viable solution for a tough problem. We saw an example of this for the air drag force; other examples of obtaining a solution with a unit analysis are the subject of several exercises below. Such a solution is always defined up to a constant dimensionless multiplier that must be determined from other considerations or from an experiment; however, merely knowing the functional form of a solution often is a great help.

Exercises

Problems with an asterisk present an extra challenge.

1. Section A.31 provides an approximate formula for the payment rate on a mortgage. Given the initial loan amount D, interest rate r, and duration of the loan T, the payment rate is given by

$$p \approx \frac{rDe^{rT}}{e^{rT} - 1}. \tag{1.11}$$

Jane used this formula to estimate payments on a mortgage that she is planning to apply for. She used the loan amount D = \$230,000, interest rate 4.5 percent (r = 0.045/year), and loan duration of T = 30 years. Using these values and equation (1.11), she estimated her payments (principal and interest) to be p = \$13,972.14, which seemed very high. The loan officer at her bank said that the correct amount of a monthly payment was p_0 = \$1,164.34. Use dimensional analysis to find a mistake in Jane's calculations.

2. Consider syrups with masses m_1, m_2, and m_3 and sugar concentrations p_1, p_2, and p_3. Sections A.16 and A.17 show that concentrations of sugar in the blend of two and three syrups are given respectively by

$$p_{12} = \frac{p_1 m_1 + p_2 m_2}{m_1 + m_2},$$
$$p_{123} = \frac{p_1 m_1 + p_2 m_2 + p_3 m_3}{m_1 + m_2 + m_3}. \tag{1.12}$$

Go through the derivations of these results and check all equations for units at each step.

3. A riverboat travels from town A to town B in time T_{AB} and from town B to town A in time T_{BA}. The time to go from town B to town A on a raft is given by

$$T_r = \frac{2T_{AB}T_{BA}}{T_{AB} - T_{BA}}. \tag{1.13}$$

Check the units for the solution of this problem in section A.2.

4. Suggest a formula for the velocity of a satellite on a circular orbit as a function of the orbit radius and the gravity acceleration. Note that orbit radius is measured in meters (m) and satellite velocity in m/s, and that the gravity acceleration is in m/s^2.

5. The law of sines and law of cosines link the lengths of the sides of a triangle and the measures of its angles (figure 1.4). Use dimensional analysis to determine which of the following formulations of these two laws are incorrect:

    a) $\sin(a\beta) = \sin(b\alpha)$

    b) $c^2 = a^2 + b^2 - 2\cos(ab\gamma)$

    c) $a\sin\beta = b\sin\alpha$

    d) $a\sin\alpha = b\sin\beta$

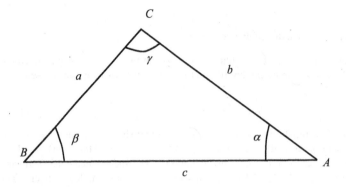

Figure 1.4
Law of sines and law of cosines: units

    e) $c^2 = a^2 + b^2 - 2\sqrt{ab}\cos\gamma$

    f) $a\sin(c + \beta) = b\sin(c + \alpha)$

    g) $c^2 = a^2 + b^2 - 2ab\cos\gamma$

    h) $a^2\sin\beta = b^2\sin\alpha$

    i) $c^3 = a^3 + b^3 - 2(ab)^{\frac{3}{2}}\cos\gamma$

6.   The waves that you may see on a beach are called gravity waves. As implied by the name, gravity acceleration $g \approx 9.8$ m/s^2 is an important parameter in modeling these waves mathematically. Two other important parameters are measured in meters:

    1. Wavelength λ, which is the distance between two consecutive crests

    2. Ocean depth h in the wave propagation area

    Using dimensional analysis, suggest two formulas for the speed of gravity waves in the ocean, one using g and λ and another one using g and h.

7.   Old clocks used a pendulum to measure time. To design a clock, we need to know the period of pendulum oscillations. A formula for this period may contain some or all of the following parameters:

    1. Gravity acceleration $g \approx 9.8$ m/s^2

    2. Length of the pendulum l, measured in meters (m)

    3. Mass of the pendulum bob M, measured in kilograms (kg)

    Suggest a formula for the period of pendulum oscillations.

8.   The radius of Archimedes's spiral increases linearly with the turn angle (figure 1.5). Use dimensional analysis to flag the incorrect formulas for the arc length of the spiral:

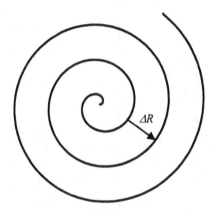

Figure 1.5
Archimedes's spiral: units

a) $\quad L = \dfrac{1}{4\pi}(\Delta R + 1)\left(\theta\sqrt{1+\theta^2} + \ln\left(\theta + \sqrt{1+\theta^2}\right)\right)$

b) $\quad L = \dfrac{1}{4\pi}\Delta R\left(\theta\sqrt{1+\theta^2} + \ln\left(\theta + \sqrt{1+\theta^2}\right)\right)$

c) $\quad L = \dfrac{1}{4\pi}\Delta R\left(\theta\sqrt{1+\theta^2} + \ln\left(\theta + \sqrt{1+\theta^2}\right) + \ln\Delta R\right)$

d) $\quad L = \dfrac{1}{4\pi}\Delta R\left(\Delta R\sqrt{1+\theta^2} + \ln\left(\theta + \sqrt{1+\theta^2}\right)\right)$

Here ΔR is the distance between the adjacent loops and θ is the total turn angle.

9. For a heavy object falling from a small height, the air resistance is relatively small, and the motion is affected primarily by the gravity (free fall). Knowing that gravity acceleration is measured in m/s^2, suggest a formula for the velocity of a falling object as a function of height h.

10. A pool must be drained for winter. There are three pumps that can be used separately or jointly. Individually, they can drain this pool in T_1, T_2, and T_3 hours. The time for draining the pool using jointly two or three pumps is given respectively by

$$
\begin{aligned}
T_{12} &= \frac{T_1 T_2}{T_1 + T_2}, \\
T_{123} &= \frac{T_1 T_2 T_3}{T_1 T_2 + T_1 T_3 + T_2 T_3}.
\end{aligned}
\tag{1.14}
$$

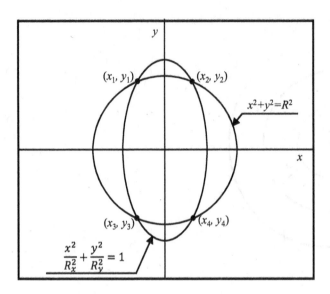

Figure 1.6
A circle and an ellipse: units

Check the units in the derivations of these solutions in sections A.18 and A.19.

11. Section A.4 solves the problem of finding intersections between a circle and an ellipse (see figure 1.6). The circle and the ellipse are given by the following equations:

$$x^2 + y^2 = R^2,$$
$$\frac{x^2}{R_x^2} + \frac{y^2}{R_y^2} = 1. \tag{1.15}$$

Assume that all coordinates are measured in meters. Use dimensional analysis for this problem to flag the wrong solutions among the following options:

a) $x = \pm \sqrt{R_x^4 \dfrac{R_y^2 - R^2}{R_y^2 - R_x^2}};$ $y = \pm \sqrt{R_y^4 \dfrac{R_x^2 - R^2}{R_x^2 - R_y^2}}$

b) $x = \dfrac{1}{4} \pm \sqrt{R_x^2 \dfrac{R_y^2 - R^2}{R_y^2 - R_x^2}};$ $y = \dfrac{1}{4} \pm \sqrt{R_y^2 \dfrac{R_x^2 - R^2}{R_x^2 - R_y^2}}$

c) $x = \pm \sqrt{R_x^2 \dfrac{R_y^2 - R^2}{R_y^2 - R_x^2}};$ $y = \pm \sqrt{R_y^2 \dfrac{R_x^2 - R^2}{R_x^2 - R_y^2}}$

d) $x = \pm \sqrt{R_x^2 \dfrac{R_y^2 - R^{-2}}{R_y^2 - R_x^2}};$ $y = \pm \sqrt{R_y^2 \dfrac{R_x^2 - R^{-2}}{R_x^2 - R_y^2}}$

12. Section A.11 deals with the sum of two scaled ratios:

$$\frac{p}{x-a} + \frac{q}{x-b} = d. \tag{1.16}$$

The solution of this equation for x is as follows:

$$x_{1,2} = \frac{(p+q) + d(a+b) \pm \sqrt{d^2(a-b)^2 + (p+q)^2 + 2d(a-b)(p-q)}}{2d}, \tag{1.17}$$

where subscripts $1, 2$ correspond to the $\pm$ signs in the right-hand side. Assume that p and q are measured in meters (m), x, a, and b are measured in seconds (s), and d is measured in m/s. Go line by line through the solution of this problem in section A.11 and check the units in each equation.

13*. The maximum angle α_{max} between a pendulum and a vertical is called the amplitude. The formula for the period of pendulum oscillations that is the solution of problem 7 is approximately valid for small amplitudes ($\alpha_{max} \ll 1$). In a case of larger amplitudes, the period should also be a function of the amplitude. For a researcher who is trying to derive such a formula, having a clue about its general structure beforehand would be extremely helpful. Identify incorrect formulas for the period of pendulum oscillations among the following options, where $F(\alpha_{max})$ is a yet unspecified transcendental function of the amplitude (refer to problem 7 for notations):

a) $\quad T = F(\alpha_{max}) \sqrt{\dfrac{l}{g}}$

b) $\quad T = F(\alpha_{max}) \sqrt{\dfrac{lM}{g}}$

c) $\quad T = F(\alpha_{max}) \sqrt{\dfrac{g}{l}}$

d) $\quad T = F(\alpha_{max}) \sqrt{lgM}$

e) $\quad T = \sqrt{\dfrac{l + F(\alpha_{max})}{g}}$

f) $\quad T = \sqrt{\dfrac{l}{g} + F(\alpha_{max})}$

14. A satellite on a circular orbit revolves around Earth every T seconds. A formula for T may contain some or all of the following parameters:

1. Gravity acceleration g at the altitude of the satellite, measured in m/s^2

2. Orbit radius R, measured in meters (m)

3. Mass of the satellite M, measured in kilograms (kg)

Suggest a formula for the orbit period of the satellite.

15. Solutions of the depressed cubic equation

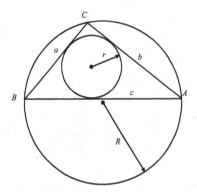

Figure 1.7
A triangle, an inscribed circle, and a circumscribed circle: units

$$x^3 + px + q = 0 \tag{1.18}$$

can be expressed through trigonometric functions (see section A.29):

$$x_k = 2\sqrt{-\frac{p}{3}} \cos\left(\frac{1}{3}\cos^{-1}\left(\frac{3q}{2p}\sqrt{-\frac{3}{p}}\right) - \frac{2\pi k}{3}\right), \tag{1.19}$$

where $k = 0, 1, 2$. Assume that x is measured in meters. Deduce the proper measurement units for parameters p and q and check the units in equation (1.19).

16. Refer to figure 1.7. The radius of the inscribed circle r can be computed using one of the following two formulas:

$$r = \frac{S}{p},$$
$$r = \frac{\sqrt{p(p-a)(p-b)(p-c)}}{p}, \tag{1.20}$$

where S is the area of the triangle, $p = (a + b + c)/2$ is its half-perimeter, and a, b, c are sides. Check the units in these two formulas.

17. We again refer to the drawing in figure 1.7. Using dimensional analysis, identify incorrect formulas for the radius of the circumscribed circle R:

a) $R = \dfrac{\sqrt[3]{a^2 b^2 c^2}}{\sqrt[4]{(a+b+c)(-a+b+c)(a-b+c)(a+b-c)}}$

b) $R = \dfrac{abc}{\sqrt{(a+b+c)(-a+b+c)(a-b+c)(a+b-c)}}$

c) $R = \dfrac{abc}{(a+b+c)(-a+b+c)(a-b+c)(a+b-c)}$

d) $R = \dfrac{abc}{\sqrt[3]{(a+b+c)(-a+b+c)(a-b+c)(a+b-c)}}$

18*. The solution of the quartic equation is even more cumbersome than that of the cubic equation. The general quartic equation is given by

$$ax^4 + bx^3 + cx^2 + dx + f = 0. \tag{1.21}$$

The root behavior is governed by the discriminant. In particular, equation (1.21) has two distinct real roots and two complex roots if the following discriminant is negative:

$$
\begin{aligned}
D = {}& 256a^3f^3 - 192a^2bdf^2 - 128a^2c^2f^2 + 144a^2cd^2f - 27a^2d^4 \\
& + 144ab^2cf^2 - 6ab^2d^2f - 80abc^2d + 18abcd^3 + 16ac^4f \\
& - 4ac^3d^2 - 27b^4f^2 + 18b^3cdf - 4b^3d^3 - 4b^2c^3f + b^2c^2d^2 .
\end{aligned}
\tag{1.22}
$$

One term in the above expression for the discriminant contains an error. Assume that f is measured in meters, and x in seconds. Identify the erroneous term using the dimensional analysis.

19*. An explosion in the air creates a spherical shock wave. The following parameters affect the shock wave propagation:

1. The energy of the explosion E, measured in kg $\cdot$ m^2/s^2

2. The density of the air ρ, measured in kg/m^3

3. Radius of the shock wave R, measured in meters (m)

Propagation of this wave away from the point of explosion is seen from the side as an expanding sphere. As the radius of the shock wave increases, the wave slows down. Using dimensional analysis, suggest a formula that computes the velocity of the shock wave (measured in m/s) as a function of the radius.

20. Section A.33 presents formulas for the linear regression algorithm. There are N data points (x_i, y_i) for variables x and y. We assume a linear model for the link between these two variables:

$$y = ax + b + R. \tag{1.23}$$

The data may not fit exactly a straight line because of measurement errors R. The linear regression algorithm states that the best estimate for parameters a and b from the data is given by the following equations:

$$a = \frac{N \sum_{i=1}^{N} x_i y_i - \sum_{i=1}^{N} x_i \cdot \sum_{i=1}^{N} y_i}{N \sum_{i=1}^{N} x_i^2 - \left(\sum_{i=1}^{N} x_i\right)^2},$$

$$b = \frac{\sum_{i=1}^{N} y_i \cdot \sum_{i=1}^{N} x_i^2 - \sum_{i=1}^{N} x_i \cdot \sum_{i=1}^{N} x_i y_i}{N \sum_{i=1}^{N} x_i^2 - \left(\sum_{i=1}^{N} x_i\right)^2}.$$

$$(1.24)$$

Determine units for a and b in these equations if x is measured in seconds (s) and y in meters (m).

2 Limiting Cases

For many centuries after Aristotle, people believed that if a moving body is not affected by any force, it must eventually slow down and stop. Galileo made an ingenious argument (adapted here) that contradicted this ancient belief. He considered a ball rolling up or down on an inclined plane. If the ball is rolling up, it decelerates. The steeper the incline, the faster the deceleration, and the shorter the distance that the ball will travel. For smaller slopes, a ball with the same starting speed would roll farther. Galileo concluded that a ball rolling on a perfectly level plane (zero slope) would roll indefinitely far.[1] This countered Aristotle's dogma and became one of the turning points in physics.

Let us look closer at Galileo's argument. He explored a system with a parameter (the incline of the plane) that can be varied at will. Then he looked at a *special value* of this parameter, which equals zero here. For this special case, the system is expected to exhibit some known behavior – that is, the lack of acceleration or deceleration. Analysis of this special case became a window into a totally new paradigm for science.

This historical example shows that it is a good idea to run analyses for various special cases in a problem. If you do this before you solve a problem, such analyses will tell you what to expect in the solution. If you do this after you have solved the problem, it is like doing an autopsy – you learn more by taking the solution apart. One powerful way to perform such analyses is checking various *limiting cases*.

Investigating limiting cases is not unlike running test scenarios on a software product. Test engineers strain different capabilities of the product; for example, they increase the load to maximum values, or check what happens if the software does not receive data from a database in time. The same is true for other industries: airplane designers must consider what happens if an engine catches fire or if the plane is struck by lightning. Unlike these examples, you do not need expensive equipment to discover interesting features of a math problem – you need just a pencil and a sheet of paper.

---

1. In practice, a ball rolling on a level plane would stop, but only because there is a force of friction.

Key Point

If a limiting case applies to the formulation of the problem, then it also applies to every step in its solution and to the final result.

Since limiting cases apply to both the original formulation of the problem and to its final result, they create a link between the starting and the ending points in your thinking. This link is largely independent from the method that was used to solve the problem and thus gives you a powerful way to check the final result. In addition, limiting cases often establish the boundaries of validity for variables and may hint at what happens beyond these boundaries. This gives you extra insight in the solution.

To evaluate your solution in a limiting case, you need to do the following:

1. Determine some *limiting or special values* for the variables in your solution where the problem shows some known or special features.
2. Check what happens when the variables take on or approach these values.
3. Check if the behavior of your solution in each limiting case is consistent with the features that you expect to see in that case.

Often you check what happens if a variable approaches zero or infinity. You may also check other values in the solution that are relevant to your specific problem. For example, if you design a system for navigating an airplane on transcontinental flights, you might want to see what happens at the poles (at the $\pm 90°$ latitudes).

Limiting cases that we study in this chapter are different from the concept of a limit that you may have studied in calculus. Typically, limiting cases explore "what if?" scenarios, where the formulation of the problem and its solution have certain known or predictable features. Often you can leverage some extraneous information to formulate a limiting case. Limits in calculus, in contrast, are based on sequences that have certain convergence properties. As such, these limits are oblivious to the special nature of a limiting case in applications. In the navigation example above, $\pm 90°$ latitudes may be investigated as limiting cases, but a mathematical limit may also exist for any other value of latitude (for example, $40°$), even though looking at these intermediate values does not shed as much light on the problem as a genuine limiting case does.

The use of limiting cases is not confined to analyzing problems you have already solved. Sometimes, a tough problem does not offer an obvious approach. In this case, it is a good tactic to take baby steps toward solving the problem before attempting the full solution. A limiting case gets your foot in the door, and you can use it to stage a major attack on the problem.

Another use of a limiting case is to become aware of the range of parameters where the problem is valid. In ancient warfare, before an army took over a town, the troops had to break through the town wall. Limiting cases, especially when they correspond to extreme values of parameters, point to a wall that defines the domain of the problem. Cracking a limiting case may produce an opening into a new and interesting area. You get a glimpse of what is behind the wall. This is often a starting point for an exploration that leads to exciting insights.

Let us now see how the magic of limiting case analysis works.

2.1   The Product of Two Linear Expressions

To illustrate how limiting cases can be checked, let's consider the following equation for x:

$$(x - a)(x - b) = d. \tag{2.1}$$

The solution of this equation is given in section A.7 and is reproduced below:

$$x_{1,2} = \frac{(a + b) \pm \sqrt{(a - b)^2 + 4d}}{2}, \tag{2.2}$$

where subscripts $1, 2$ correspond to the $\pm$ signs in the right-hand side. Below we consider three limiting cases and check both the original problem and its solution for each limiting case:

1. The case of $d = 0$. The right-hand side of the original equation (2.1) becomes zero. The left-hand side is a product of two factors; in order for it to be zero, one or both of them must be zero. Therefore, we expect the original equation to have two roots, $x_1 = a$ and $x_2 = b$. If we turn to the solution, we should obtain the same two values for x for this limiting case. Indeed, setting $d = 0$ transforms equation (2.2) into the following:[2]

$$
\begin{aligned}
x_{1,2} &= \frac{(a + b) \pm \sqrt{(a - b)^2}}{2} \\
&= \frac{(a + b) \pm (a - b)}{2},
\end{aligned}
\tag{2.3}
$$

   which does produce $x_1 = a$ and $x_2 = b$.

2. The case of $a = b$. The left-hand side of the original equation becomes a complete square:

---

2. To be precise, $\sqrt{(a - b)^2} = |a - b|$, which is not necessarily equal to $(a - b)$. However, the $\pm$ signs make applying the absolute value unnecessary.

$$(x - a)^2 = d. \tag{2.4}$$

Extracting the square root from both sides produces the following solution for x:

$$x_{1,2} = a \pm \sqrt{d}, \tag{2.5}$$

where subscripts $1, 2$ correspond to the $\pm$ signs in the right-hand side.

Let us now check this limiting case for equation (2.2). We observe that the term $(a-b)^2$ becomes zero, and the term $(a+b)$ is now equal to $2a$, yielding a simplification:

$$x_{1,2} = \frac{2a \pm \sqrt{4d}}{2}. \tag{2.6}$$

From this, we do obtain $x_{1,2} = a \pm \sqrt{d}$, as expected.

3. Finally, we explore the limiting case for $a = -b$. In the original equation, we get

$$(x - a)(x + a) = d. \tag{2.7}$$

We expand $(x - a)(x + a) = x^2 - a^2$ and solve the last equation for x:

$$x_{1,2} = \pm \sqrt{a^2 + d}. \tag{2.8}$$

For equation (2.2), substitution of $a = -b$ removes the $a + b$ term in the numerator, and $(a - b)$ can be now replaced by $2a$. The resulting equation does produce $x_{1,2} = \pm \sqrt{a^2 + d}$.

As we see, all three limiting cases hold. This is a confirmation for the validity of our solution, even if not a definitive one.

2.2 The Two Hikers Problem

Consider two hikers that walk toward each other (figure 2.1). One hiker maintains speed V_1, and another maintains speed V_2. The total length of the trail is D. Section A.1 shows that hikers 1 and 2 will travel the following distances:

$$D_1 = \frac{DV_1}{V_1 + V_2},$$
$$D_2 = \frac{DV_2}{V_1 + V_2}. \tag{2.9}$$

This problem has two limiting cases:

1. What happens if V_1 is very small compared to V_2? Then from equation (2.9) follows that D_1 is small compared to D (denoted as $D_1 \ll D$) and that $D_2 \approx D$. This makes sense: most of the entire distance will be covered by the faster hiker.

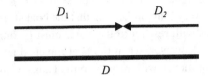

Figure 2.1
Two hikers on a trail: limiting cases

2. The two distances must sum to the total distance: $D_1 + D_2 = D$. A direct substitution shows that this condition holds:

$$D_1 + D_2 = \frac{DV_1}{V_1 + V_2} + \frac{DV_2}{V_1 + V_2}$$
$$= \frac{DV_1 + DV_2}{V_1 + V_2} \quad\quad (2.10)$$
$$= D.$$

   Both limiting cases hold, as expected. This increases our confidence in the validity of the solution.

2.3 The Riverboat Problem

Assume that a riverboat travels upstream from town A to town B in time T_{AB} and downstream from town B to town A in time T_{BA}. Section A.2 shows that a raft trip from town B to town A can be completed in time T_r, given by the following equation:

$$T_r = \frac{2T_{AB}T_{BA}}{T_{AB} - T_{BA}}. \quad\quad (2.11)$$

   For this problem, we consider the following three limiting cases:

1. What happens if $T_{AB} \approx T_{BA}$? This would mean that it takes approximately the same time to travel by a boat upstream as it does to travel downstream. In this case, the effect of the river current on the boat speed is negligible, which tells us that the river current is weak (compared to the boat speed). If the river current is weak, a raft trip

will take a long time. This is consistent with the solution (2.11): the denominator is small, making the travel time large.

2. What happens if T_{AB} is much larger than T_{BA} (mathematically expressed as $T_{AB} \gg T_{BA}$)? In this case, the travel time upstream is very long, indicating that the boat struggles to go up the river. The speed of the boat with respect to the water must be only slightly higher than the speed of the current. However, when the boat travels downstream, these two speeds add up, and the boat moves with respect to the banks of the river approximately at twice the speed of the current. A raft would travel at the speed of the current, or at a half of the boat speed. Correspondingly, it will take approximately twice the time that the boat takes. Indeed, if $T_{AB} \gg T_{BA}$, we get

$$T_r = \frac{2T_{AB}T_{BA}}{T_{AB} - T_{BA}} \approx 2T_{BA}. \tag{2.12}$$

3. For the downstream trip, we expect the riverboat to be faster than the raft, which means that $T_r - T_{BA} > 0$. Indeed, using equation (2.11), we get

$$
\begin{aligned}
T_r - T_{BA} &= \frac{2T_{AB}T_{BA}}{T_{AB} - T_{BA}} - T_{BA} \\
&= \frac{2T_{AB}T_{BA} - T_{AB}T_{BA} + T_{BA}^2}{T_{AB} - T_{BA}} \\
&= T_{BA}\frac{T_{AB} + T_{BA}}{T_{AB} - T_{BA}} \\
&> 0.
\end{aligned}
\tag{2.13}
$$

This result is conditioned on the current convention for T_{AB}, T_{BA}: the former is upstream, the latter is downstream, and $T_{AB} > T_{BA}$. You may also check that no similar inequality can be proven for $T_r - T_{AB}$ because there is no meaningful comparison to be made for the travel times of the boat going upstream and the raft going downstream.

In the beginning of this chapter I noted that limiting cases can provide a good way to identify erroneous solutions. Suppose we made a mistake in solving this problem, and instead of equation (2.11) we got the following (wrong) answer:

$$\text{🐢} \quad T_r' = \frac{2T_{AB}T_{BA}}{T_{AB} + T_{BA}} \quad \text{🐟}. \tag{2.14}$$

Limiting case 1 would immediately tell us that equation (2.14) is incorrect: the value of T_r' does not go to infinity when $T_{AB} \approx T_{BA}$. However, even if a solution passes this particular check, it does not guarantee that it is correct. For example, suppose that instead of flipping a sign in the denominator, we have lost a multiplier of 2 in the solution to get another wrong answer:

$$\text{🐸} \quad T_r'' = \frac{T_{AB}T_{BA}}{T_{AB} - T_{BA}} \quad \text{🦈} . \tag{2.15}$$

This solution has the desired feature: the value of T_r'' goes to infinity as T_{AB} approaches T_{BA}, but it is still wrong. Yet, this error can be caught by our second limiting case $T_{AB} \gg T_{BA}$. Indeed, we expect to have $T_r \approx 2T_{BA}$, which does not work for equation (2.15). We see that applying both limiting cases would point us to various types of possible errors in the solution of this problem. Luckily, the combination of these two limiting cases would catch both errors.

While checking for errors is helpful, there is another benefit from considering limiting cases. Arguably, we've got a better understanding of the problem and the trade space of parameters. For real-life tasks, such as designing a new system or analyzing experimental data, this is very important. We always want to see how the real system "behaves" for different setups. The riverboat problem, while simple, gives you a taste of how limiting cases help us explore the range of possibilities for a problem.

2.4 The Quadratic Equation

This example of limiting case analysis deals with the ubiquitous quadratic equation:

$$ax^2 + bx + c = 0. \tag{2.16}$$

Its two solutions are given by the standard quadratic formula:

$$x_{1,2} = \frac{-b \pm \sqrt{b^2 - 4ac}}{2a}, \tag{2.17}$$

where subscripts $1, 2$ correspond to the $\pm$ signs in the right-hand side. Let us look at two limiting cases here:

1. Consider both solutions in the limiting case $c = 0$. The equation takes the form

$$ax^2 + bx = 0. \tag{2.18}$$

This is equivalent to

$$x(ax + b) = 0. \tag{2.19}$$

We see that one of the roots is zero ($x_1 = 0$), and (assuming $a \neq 0$) the second one is given by $x_2 = -b/a$. If this is true for equation (2.16), it must be true for its solutions (2.17) as well. When we plug $c = 0$ in the right-hand side of equation (2.17), we get

$$x_{1,2} = \frac{-b \pm |b|}{2a}. \tag{2.20}$$

This yields the two desired solutions:

$$x_1 = 0,$$
$$x_2 = \frac{-b}{a}. \tag{2.21}$$

This limiting case holds.

While this example is trivial, it shows the key feature of limiting case analysis. We can predict what happens in a limiting case from the original formulation of the problem, and our solution must conform to this prediction. This establishes a link between the formulation of the problem and its solution. Had we made an error in the derivation of the final result, that link could have been broken. If a limiting case does not hold, we know that the solution is flawed. On the other hand, each valid limiting case increases our confidence in the validity of the final result.

2. The next limiting case is less trivial. Let us see what happens if coefficient a in equation (2.16) approaches zero for some fixed values of b and c. We know that a quadratic equation with real coefficients has either two real roots or two complex roots. For example, it has two roots if $a = 10^{-100}$, $b = 2$, and $c = 1$. However, if the value of a changes by a tiny amount and is set to zero, the quadratic term disappears, the equation becomes linear, and now it has just one root. Where did the second root go?

We may recall that both roots of a quadratic equation become complex if the discriminant is negative, and thus they disappear from the real axis. However, for real-valued coefficients, this may only apply to *both* roots at the same time, not one of them. The disappearance of one root of equation (2.16) cannot be explained by assuming that one root becomes complex. Moreover, we can see that the discriminant of this equation is nonnegative for small enough values of $|a|$, and therefore both roots must be real.

The "disappearing root question" illustrates how a limiting case provides additional insight in the problem. Let us turn to equation (2.17) and see what happens if $a \to 0$. In the numerator, the value of the square root approaches $|b|$, which means that the value of the numerator is close to either $-2b$ or zero, depending on the selection of the sign. Separately, the denominator approaches zero. If the numerator is close to $-2b$ and the denominator is close to zero, the corresponding root of the quadratic equation has a very large absolute value. Therefore, as $a \to 0$, this root goes to infinity. Indeed, the infinity is the only "cliff" on a real line where a root can disappear.

Figure 2.2 shows the behavior of roots of a quadratic equation as a function of a for $b = 20, c = 10$. For small $|a|$, one of the roots does go to infinity, while the other one stays put. For larger values of a, the values of both roots converge: at $a = 10$, where the discriminant of this equation is zero, the two roots have the same value. There are no real-valued roots for $a > 10$.

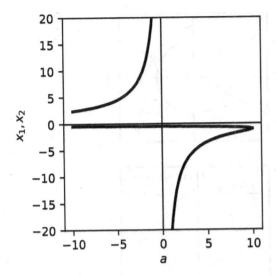

Figure 2.2
The quadratic equation: limiting cases

Figure 2.2 also shows what happens with the second root when $a \to 0$: this root stays finite. Of course, if we simply set $a = 0$, the quadratic equation becomes linear, and its solution is given by

$$x = \frac{-c}{b}.$$ (2.22)

Can we get this limiting case from equation (2.17)? The computation for $a \to 0$ is not trivial. The standard way to do that involves calculus, which is beyond the scope of this book. When we study symmetry later, we will learn how to find the value of that root from equation (2.17) for $a \to 0$ without the use of calculus (see section 3.12).

2.5   The Intersections between a Circle and a Straight Line

A circle and a straight line are given by the following equations:

$$x^2 + y^2 = R^2,$$
$$y = px + q.$$ (2.23)

Section A.3 shows that the horizontal coordinates of the intersections of a circle and a line (if any) are given by

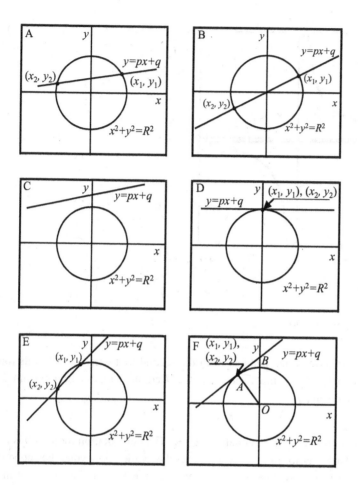

Figure 2.3
A circle and a straight line: limiting cases

$$x_{1,2} = \frac{-pq \pm \sqrt{(1 + p^2) R^2 - q^2}}{1 + p^2},$$

(2.24)

where subscripts $1, 2$ correspond to the $\pm$ signs in the right-hand side. An example of the geometry of this problem is shown in figure 2.3A. Below we analyze limiting cases for equations (2.23) and their solution (2.24).

It is helpful to recall the meaning of various parameters in equations (2.23). The radius of the circle is R, the straight line has slope p, and it crosses the vertical axis at $y = q$. There are several limiting cases that we explore for this problem:

1. If $q = 0$, the discriminant $(1 + p^2)R^2 - q^2 = (1 + p^2)R^2$ in equation (2.24) is always nonnegative; therefore, two real solutions exist. This corresponds to the straight line that goes through the center of the circle (figure 2.3B). Such a line cannot "miss" the circle, ensuring that intersection points do exist. This limiting case holds.

2. On the other hand, if $q \to \pm\infty$ and other parameters are finite, the straight line moves up or down by a large amount and will not intersect the circle (figure 2.3C). We expect the equations to have no roots, which happens if the discriminant is negative. Indeed, if the absolute value of the second term $-q^2$ becomes sufficiently large, the discriminant $(1 + p^2)R^2 - q^2$ will turn negative, and there will be no real roots.

3. Assume that the slope of the straight line p is zero and that $q = R$. On the figure, this would mean that the straight line is tangent to the circle at the top (see figure 2.3D). The two intersections merge into one point, and we must get $x_1 = x_2 = 0$. We know that the condition for two roots of a quadratic equation to merge is the discriminant being equal to zero. Indeed, if $p = 0$, $q = R$, we get $(1 + p^2)R^2 - q^2 = 0$. This limiting case holds. By a direct substitution, we can check that this case yields $x_{1,2} = 0$. Similarly, we get $x_{1,2} = 0$ if $p = 0$ and $q = -R$; this corresponds to the straight line being tangent to the circle at the bottom.

4. Suppose that $q > R$, which means that the straight line crosses the vertical axis above the circle. We also assume that the line slope p is sufficiently large to produce intersections with the circle (see figure 2.3E). In this case, both intersections are on the same side of the vertical axis. Does our solution exhibit this property?

   For this property to be true, the value of the square root in the solution should be less than the value of $|pq|$. In this case, the addition or subtraction of the square root, while altering the value of the solution, will not flip its sign, causing both roots $x_{1,2}$ to be on the same side of the vertical axis. Mathematically, this is expressed as

$$|pq| > \sqrt{(1 + p^2)R^2 - q^2}. \tag{2.25}$$

We square this equation to get

$$p^2q^2 > \left(1 + p^2\right)R^2 - q^2. \tag{2.26}$$

Rearranging the terms yields

$$\left(1 + p^2\right)q^2 > \left(1 + p^2\right)R^2. \tag{2.27}$$

If $q > R$, the inequality does hold. This special case also works. Note that it works also for $q < -R$, which corresponds to the straight line intersecting the vertical axis below the circle.

The boundary of applicability for this case is when the line crosses the circle right at the top or at the bottom, which corresponds to $q = \pm R$. We expect one of the roots for x to be zero in this case. This particular value separates the scenario when both roots have the same sign from the scenario when they have different signs. Substituting $q = R$ in equation (2.24) yields

$$
\begin{aligned}
x_{1,2} &= \frac{-pR \pm \sqrt{(1 + p^2) R^2 - R^2}}{1 + p^2} \\
&= \frac{-pR \pm |pR|}{1 + p^2},
\end{aligned}
\tag{2.28}
$$

where we do get a zero for one choice of the sign. Case $q = -R$ is analogous.

5. We have looked at the case when the straight line is horizontal and is tangent to the circle at the top or at the bottom. What happens if it is tangent to the circle elsewhere (figure 2.3F)? In this case, roots for x should have the same value: $x_1 = x_2$. Mathematically, this means that the discriminant of the quadratic equation is zero (see equation (2.24)). Therefore, we must have $(1 + p^2)R^2 = q^2$. Proving this from triangle OAB in figure 2.3F is left as an exercise for the reader.

We see that limiting cases unearthed a treasure of information about the problem. They helped us both check the solution and understand it better. Moreover, applying each limiting case did not require any extraordinary effort.

2.6   The Sum of Two Ratios

Consider the following equation for x:

$$
\frac{1}{x - a} + \frac{1}{x - b} = d.
\tag{2.29}
$$

Its solution is as follows (see section A.10 for details):

$$
x_{1,2} = \frac{2 + d(a + b) \pm \sqrt{d^2(a - b)^2 + 4}}{2d}.
\tag{2.30}
$$

Several limiting cases show what happens with the roots when the parameters take some special values:

1. Let's see what happens with the original equation if d is very large. In equation (2.29), the right-hand side is very large, which means that at least one of the two terms in the left-hand side must be large. Since the numerators in both are equal to one, at least one of the denominators must be small. Then $x \approx a$ or $x \approx b$ (or both, if it happens that

$a \approx b$). Do these conditions hold for the solution? If we assume that d is very large in equation (2.30), some terms in the numerator become dominant:

$$d^2(a-b)^2 \gg 4,$$
$$|d(a+b)| \gg 2. \tag{2.31}$$

If we neglect the smaller terms, equation (2.30) takes the form[3]

$$x_{1,2} \approx \frac{d(a+b) \pm \sqrt{d^2(a-b)^2}}{2d} = \frac{(a+b) \pm (a-b)}{2}. \tag{2.32}$$

The right-hand side does compute to

$$x_1 \approx a,$$
$$x_2 \approx b. \tag{2.33}$$

This limiting case holds.

2. Let's now see what happens if d is close to zero for some fixed and finite values of a and b. In equation (2.29) the right-hand side is close to zero, which means that either both terms in the left-hand side are close to zero or they nearly cancel each other.

(a) In the first option, the absolute values of $1/(x-a)$ and $1/(x-b)$ in equation (2.29) are small. Since the numerators there are equal to one, the absolute values of both denominators must be large. Since a and b are fixed, the absolute value of x must be large. More precisely, if $|x|$ is large compared to $|a|$ and $|b|$, we can neglect these two parameters; then equation (2.29) is in the form

$$\frac{1}{x_1} + \frac{1}{x_1} \approx d. \tag{2.34}$$

This yields

$$x_1 \approx \frac{2}{d}. \tag{2.35}$$

For a small value of d, we do get a large value of x_1.

The essence of our analysis is to connect the features of the original formulation of the problem with those of the solution. We have considered the effect of a small $|d|$ on the original equation; now we turn to solution (2.30). Indeed, if we assume that $|d|$ is small, we can neglect terms with d in the numerator of equation (2.30). Note that we cannot neglect d in the denominator because it is the only term there:

---

3. To be precise, $\sqrt{d^2(a-b)^2} = |d(a-b)|$, which is not necessarily equal to $d(a-b)$. However, the $\pm$ signs make applying the absolute value unnecessary.

$$x_{1,2} = \frac{2 + d(a+b) \pm \sqrt{d^2(a-b)^2 + 4}}{2d}$$

$$\approx \frac{2 \pm \sqrt{4}}{2d}.$$

(2.36)

For the plus sign, we do get

$$x_1 \approx \frac{2}{d}.$$

(2.37)

This limiting case holds.

(b) In the second option, the two terms in the left-hand side of (2.29) nearly cancel out. We require that

$$\frac{1}{x_2 - a} \approx -\frac{1}{x_2 - b}.$$

(2.38)

This is equivalent to

$$(x_2 - a) \approx -(x_2 - b),$$

(2.39)

which yields

$$x_2 \approx \frac{a+b}{2}.$$

(2.40)

Checking this limiting case in solution (2.30) without using calculus is tricky. One way to do it is to compute the product of the two roots. We substitute both values from solution (2.30) to get

$$x_1 x_2 = \frac{2 + d(a+b) + \sqrt{d^2(a-b)^2 + 4}}{2d}$$

$$\times \frac{2 + d(a+b) - \sqrt{d^2(a-b)^2 + 4}}{2d}.$$

(2.41)

After some algebra, we get

$$x_1 x_2 = \frac{4 + 4d(a+b) + d^2(a+b)^2 - 4 - d^2(a-b)^2}{4d^2}$$

$$= ab + \frac{a+b}{d}.$$

(2.42)

For small values of d, we can neglect the product ab, and the last equation is approximated as

$$x_1 x_2 \approx \frac{a+b}{d}.$$

(2.43)

Since we already know that $x_1 \approx 2/d$ (see equation (2.35)), we conclude that

$$x_2 \approx \frac{a+b}{2}. \tag{2.44}$$

This limiting case also holds (compare with equation (2.40)), though checking it has required some work.

3. Finally, consider the case where, for a particular application, d is expressed through some other parameter c:

$$d = \frac{1}{c-a} + \frac{1}{c-b}. \tag{2.45}$$

The solution of equation (2.29) with this special value of d in principle is not different from the original and would produce two roots given by equation (2.30), where d is in the form of equation (2.45). On the surface, we expect these roots to be expressed through radicals, as they are in equation (2.30). However, there is a twist here; that is the subject of the following argument.

Substituting d from equation (2.45) in the original equation (2.29) yields

$$\frac{1}{x-a} + \frac{1}{x-b} = \frac{1}{c-a} + \frac{1}{c-b}. \tag{2.46}$$

If we inspect both sides of this equation, we see that $x = c$ satisfies this equation; hence, one of the roots must be equal to c. As discussed above, the general formula is still given by equation (2.30), where d is expressed through c. Therefore, if we substitute the special value of d from equation (2.45) in solution (2.30) and simplify the resulting expression, we should get just $x = c$ for one of the $\pm$ signs. Note that solution (2.30) will no longer have a radical for that root. However, it is difficult to see how the radical can disappear, unless the expression in the radical happens to be a complete square.

If we do get a complete square in the radical, that would affect the result for both roots, not just for the root $x = c$. Therefore, there is a good chance that the second root will be expressed through parameters a, b, c using only the basic arithmetic operations (that is, $+, -, \times, /$) and without a radical. This is an interesting conjecture that is worth checking out.

Indeed, a messy algebraic derivation shows that for d given by equation (2.45), the discriminant in equation (2.30) does become a full square:

$$d^2(a-b)^2 + 4 = \frac{\left(a^2 - 2ac + b^2 - 2bc + 2c^2\right)^2}{(c-a)^2(c-b)^2}. \tag{2.47}$$

Substituting this expression for the discriminant in solution (2.30) does produce $x_1 = c$ for the plus sign. For the minus sign, we get

$$x_2 = \frac{c(a+b) - (a^2 + b^2)}{2c - (a+b)}.$$

(2.48)

In this special case for d we arrive at a much simpler expression than we would expect from the raw formulas with a radical, and the simplification became possible because we have noticed that for d given by equation (2.45), the radicals in the solution should disappear.

This limiting case is probably not a good means to check the validity of equation (2.30) because checking it has required a lot of work. This limiting case is presented here to show how it can help make a prediction about the form of the final result. Without this analysis, detecting a complete square in the discriminant for d given by equation (2.45) and getting a simplified result would be a formidable task!

2.7   The Sum of Two Scaled Ratios

This section illustrates a common way of dealing with complex problems that is based on using limiting cases:

1. Formulate another problem that is easier to solve and that happens to be a limiting case of the problem in hand.
2. Solve the easier, limiting-case problem first.
3. Extend your solution to the original problem, while keeping an eye on the limiting case at every step.

This technique gives two benefits: first, we approach a complex problem in a gradual, deliberate way; second, we check every step of solving the harder problem. This is a great way to prevent errors in our solution.

Below we compare two problems to illustrate this technique. One is the already familiar problem for the sum of two ratios that we explored in section 2.6. We consider the following equation:

$$\frac{1}{x-a} + \frac{1}{x-b} = d.$$

(2.49)

Its solutions are given by

$$x_{1,2} = \frac{2 + d(a+b) \pm \sqrt{d^2(a-b)^2 + 4}}{2d}.$$

(2.50)

The second problem is for the sum of two *scaled* ratios (see section A.11 for details). We solve the following equation for x:

$$\frac{p}{x-a} + \frac{q}{x-b} = d.$$

(2.51)

The solution is as follows:

$$x_{1,2} = \frac{(p+q) + d(a+b) \pm \sqrt{d^2(a-b)^2 + (p+q)^2 + 2d(a-b)(p-q)}}{2d}, \qquad (2.52)$$

where subscripts 1, 2 correspond to $\pm$ signs in the right-hand side.

As you can see, these problems are related. Specifically, equation (2.49) is a limiting case of equation (2.51) for $p \to 1, q \to 1$. Table 2.1 illustrates the solutions of equations (2.49) and (2.51) when we look at them step by step. Following derivations in sections A.10 and A.11, table 2.1 distills the intermediate steps for both problems in separate columns.

The validity of the limiting case $p \to 1, q \to 1$ is maintained throughout the entire derivation. At every step, you can take an equation from the second column of the table, apply conditions $p \to 1, q \to 1$, and obtain the corresponding equation in the first column. (Doing that is the subject of an exercise at the end of this chapter.)

Table 2.1
Parallel solutions of two related problems

Limiting case	Full equation	Step description
$\dfrac{1}{x-a} + \dfrac{1}{x-b} = d$	$\dfrac{p}{x-a} + \dfrac{q}{x-b} = d$	Original equation
$(x-b) + (x-a) =$ $d(x-a)(x-b)$	$p(x-b) + q(x-a) =$ $d(x-a)(x-b)$	Multiply by denominators
$dx^2 - x(2 + d(a+b)) +$ $(a+b) + dab = 0$	$dx^2 - x((p+q) + d(a+b)) +$ $(qa + pb) + dab = 0$	Collect terms
$D = (2 + d(a+b))^2 -$ $4d((a+b) + dab)$	$D = ((p+q) + d(a+b))^2 -$ $4d((qa + pb) + dab)$	Discriminant
$D = d^2(a-b)^2 + 4$	$D = d^2(a-b)^2 + (p+q)^2 +$ $2d(a-b)(p-q)$	Simplified discriminant
$x_{1,2} = \dfrac{2 + d(a+b)}{2d} \pm$ $\dfrac{\sqrt{d^2(a-b)^2 + 4}}{2d}$	$x_{1,2} = \dfrac{(p+q) + d(a+b)}{2d} \pm$ $\dfrac{\sqrt{d^2(a-b)^2 + (p+q)^2 + 2d(a-b)(p-q)}}{2d}$	Final result

> **Key Point**
>
> When dealing with a difficult problem, try to solve it for a simpler limiting case first. Then use that result to build up a solution for the original problem.

The two problems in this section are not overly complex, and their solutions are not overly long. However, a real-life application may require a long and cumbersome derivation. In such a case, finding a simpler limiting-case version of a problem helps immensely.

2.8   The Sum or Difference of Two Radicals

Sections A.20 and A.21 present solutions for the following equation:

$$\sqrt{a+x} \pm \sqrt{a-x} = b, \tag{2.53}$$

where $a > 0$.

For the plus sign in equation (2.53) and $b > 0$, the solutions are given by

$$x_{1,2} = \pm \sqrt{a^2 - \left(\frac{b^2}{2} - a\right)^2}, \tag{2.54}$$

where subscripts $1, 2$ correspond to the $\pm$ signs in the right-hand side of equation (2.54).

For the minus sign in equation (2.53), the solutions are given by

$$x = \sqrt{a^2 - \left(\frac{b^2}{2} - a\right)^2}, \quad \text{if} \quad b \geq 0,$$

$$x = -\sqrt{a^2 - \left(\frac{b^2}{2} - a\right)^2}, \quad \text{if} \quad b < 0. \tag{2.55}$$

A quick look at equation (2.54) for x_1 and the first equation of (2.55) shows that roots for x have the same functional form if $b > 0$. This is puzzling because they have to satisfy two different forms of equation (2.53) containing different signs for $\sqrt{a-x}$. The same value of x may not satisfy both versions, except for a single case when $\sqrt{a-x} = 0$, when the sign for that term does not matter. Below we use limiting cases to clarify this issue.

The clue to the puzzle is given by that extreme case when the same value of x can satisfy equations with different signs. Specifically, we look at the limiting case

$$\sqrt{a-x} = 0. \tag{2.56}$$

In this case, $x = a$, and a direct substitution in equation (2.53) shows that this requires

$$b = \sqrt{2a}. \tag{2.57}$$

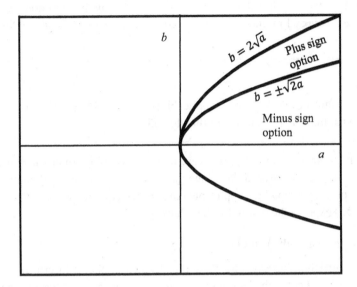

Figure 2.4
Parameter domains for the sum and for the difference of two radicals

(Note that for the expressions under the square roots in equation (2.53) to be nonnegative, we must require $-a \leq x \leq a$; therefore, $x = a$ is indeed a limiting case for the domain of allowed values of x.)

Equation (2.57) is a condition on parameters a and b. We shall see that it separates two distinct domains for these parameters; these domains overlap nowhere, except at one special case: equation (2.57). Expression $b \leq \sqrt{2a}$ applies for the plus sign in equation (2.53), and $b \geq \sqrt{2a}$ applies for the minus sign. Indeed, squaring equation (2.53) produces

$$2a \pm 2\sqrt{a^2 - x^2} = b^2, \tag{2.58}$$

or

$$b^2 - 2a = \pm 2\sqrt{a^2 - x^2}. \tag{2.59}$$

Since $\sqrt{a^2 - x^2} \geq 0$, we conclude that $b^2 \geq 2a$ for the plus sign and $b^2 \leq 2a$ for the minus sign. The boundary value $b = \sqrt{2a}$ in equation (2.57) separates these two ranges. On top of that, equation (2.58) mandates that $b \leq 2\sqrt{a}$ for the plus sign option because the left-hand side does not exceed $4a$. The domains of validity for equation (2.53) are shown in figure 2.4. We see that these domains overlap at one line only: for $b = \sqrt{2a}$.

We started going down the rabbit hole of why the same value of x satisfies two seemingly incompatible equations. Turns out there was no conflict here: the two equations apply to different ranges of parameters a and b that overlap only at the mutually valid boundary $b = \sqrt{2}a$. This boundary, being a limiting case for each equation, gave us a clue for solving the puzzle.

Key Point

Limiting cases may point to the domain of applicability for a problem. Often it is helpful to explore what may happen outside of that domain.

A takeaway from this problem is that a limiting case may define a range of applicability for a problem. Moreover, peeking beyond this range may give you an idea about another problem. Specifically, peeking beyond the applicable range of parameters for the plus sign option in equation (2.53) exposes the option for the minus sign.

2.9   A Circle Inscribed in a Right Triangle

In this section, we consider limiting cases for the radius of a circle that is inscribed in a right triangle (see figure 2.5). One of the legs has length a, and the measure of the adjacent angle is α. According to section A.27, the radius of the inscribed circle is

$$R = \frac{a \sin \alpha}{1 + \sin \alpha + \cos \alpha}. \tag{2.60}$$

Let's consider limiting cases for this problem. There are two parameters to deal with: a and α. When we analyze limiting cases for one of these parameters, we will assume that the other parameter has a value that does not correspond to a limiting case. This will allow

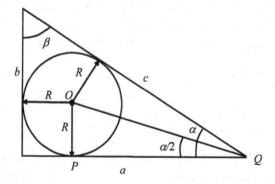

Figure 2.5
A circle inscribed in a right triangle: limiting cases

us to disentangle the effects of different limiting cases. This is a common technique to use. We will consider four limiting cases, each affecting the value of a or α, but not both:

1. The length of the leg is small: $a \to 0$ and a "normal" value of α, that is, not close to limiting values 0 and $\pi/2$. If we have a small triangle, we only can inscribe a small circle in it (figure 2.5 would just become proportionally smaller). This is also evident from equation (2.60).

2. The length of the leg is large: $a \to \infty$. This is the flip side of the previous limiting case, with a similar argument. We expect R to be correspondingly large, which is indeed predicted by equation (2.60). The triangle in figure 2.5 would just become proportionally larger.

3. The angle is small: $\alpha \to 0$. This time, we hold the value of a at some fixed level. This case is illustrated in figure 2.6A. The triangle is squeezed, making the circle small. Indeed, in equation (2.60) the numerator approaches zero, while the denominator remains finite. The radius of the circle approaches zero as well.

4. The angle is close to the maximum possible value: $\alpha \approx \pi/2$. Again, it is instructive to hold the value of the other parameter a at some fixed level. Figure 2.6B shows what happens in this case. The triangle is squeezed in a different direction, but for a fixed length of a, the other leg becomes correspondingly longer, and the circle does *not* become small. We can see that the other leg becomes almost parallel to the hypotenuse.

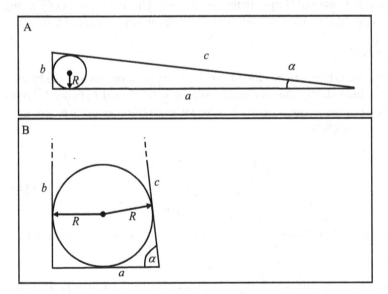

Figure 2.6
Small and large angles for a right triangle with an inscribed circle

From the figure you can see that the diameter of the circle must be close to a, which is equivalent to $R \approx a/2$.

Let us see if this is the case for our solution. In equation (2.60) we get $\sin \alpha \approx 1$, $\cos \alpha \approx 0$, which does yield the condition $R \approx a/2$. This limiting case also holds. Note that in this limiting case the other angle of the triangle $\beta = \pi/2 - \alpha$ becomes small. Therefore, this limiting case looks like another application of the previous limiting case, but applied to the other nonright angle of the triangle. Yet, the radius of the circle does not go to zero here. This is because in both limiting cases we kept a constant, and this leg is adjacent to the small angle in limiting case 3 and is opposite to the small angle in limiting case 4.

Key Point

When applying limiting cases to a parameter, it is common to hold values of other parameters at some nominal values.

The takeaway from this analysis is that it is often helpful to separate the effects of different limiting cases in a way that facilitates your analysis. Limiting cases $\alpha \to 0$ and $\alpha \approx \pi/2$, when applied for fixed values of a, produce very different outcomes: triangles become "small" in the first case and "large" in the second case. This is an artifact of keeping a constant in each case. Yet, both cases work well as checks on our solution (2.60).

2.10 Draining a Pool

Consider a pool that needs to be drained for winter. There are three pumps that can be used separately or jointly. Individually, they can drain this pool in T_1, T_2, and T_3 hours. Sections A.18 and A.19 show that the amounts of time required to drain this pool using two or three pumps are given respectively by

$$T_{12} = \frac{T_1 T_2}{T_1 + T_2},$$
$$T_{123} = \frac{T_1 T_2 T_3}{T_1 T_2 + T_1 T_3 + T_2 T_3}. \tag{2.61}$$

These problems feature four limiting cases that can be used to identify errors, if any, in the solutions:

1. First, we look at a case when one pump is much more powerful than another one. Suppose we have two pumps and pump 1 is many times more powerful than pump 2. This would mean that if the pumps are used individually, pump 1 will drain the pool much faster than pump 2:

$$T_1 \ll T_2. \tag{2.62}$$

This would also mean that if these two pumps are used together, pump 2 will not make much of a difference compared to using pump 1 alone. We should expect that

$$T_{12} \approx T_1. \tag{2.63}$$

Indeed, if we use $T_1 \ll T_2$ in the first equation of (2.61), we can neglect T_1 in the denominator and get the desired result $T_{12} \approx T_1$.

The same argument applies to the case of three pumps, if one of them (for example, pump 1) completely overpowers the other two. If $T_1 \ll T_2, T_3$, then the expression for T_{123} in equations (2.61) is approximated as follows:

$$
\begin{aligned}
T_{123} &= \frac{T_1 T_2 T_3}{T_1 T_2 + T_1 T_3 + T_2 T_3} \\
&\approx \frac{T_1 T_2 T_3}{T_2 T_3} \\
&= T_1.
\end{aligned}
\tag{2.64}
$$

2. Next, we look at a case of three pumps when one of them (for example, pump 3) is very weak compared to the other two. The time required to drain the pool using just that weak pump must be very long:

$$T_3 \gg T_1, T_2. \tag{2.65}$$

When all three pumps are working together, this third pump does not matter. The problem for the three-pump case becomes approximately equivalent to the problem of draining the pool with pumps 1 and 2 only. We should be able to see that case in our equations as well. We start from the second equation in (2.61) and assume that $T_3 \gg T_1, T_2$:

$$
\begin{aligned}
T_{123} &= \frac{T_1 T_2 T_3}{T_1 T_2 + T_1 T_3 + T_2 T_3} \\
&\approx \frac{T_1 T_2 T_3}{T_1 T_3 + T_2 T_3} \\
&= \frac{T_1 T_2}{T_1 + T_2}.
\end{aligned}
\tag{2.66}
$$

Indeed, this case reduces to the two-pump problem (see the first equation in system (2.61))!

3. Next, we consider the problem with two pumps of equal power. In this case, we intuitively expect the pool to be drained in half of the time when using both of them. If we use $T_1 = T_2$ in the first equation of (2.61), we do get $T_{12} = T_1/2 = T_2/2$. Similarly, if we assume that all three pumps in the second equation of (2.61) have equal power, we get the expected result: using them jointly drains the pool in a third of the time.

4. The next check is not a limiting case as such, but it is a good way to validate the solution. We expect that using more pumps should shorten the drain time from the scenario with fewer pumps. This means that solutions (2.61) should obey the following inequalities:

$$T_{12} < T_i,$$
$$T_{123} < T_j, \qquad\qquad (2.67)$$
$$T_{123} < T_{12},$$

where $i = 1, 2$ and $j = 1, 2, 3$. To prove inequalities (2.67), we compute the difference between the left- and the right-hand side in each of them. For a comparison with pump 1 (that is, for $i = j = 1$) we get

$$
\begin{aligned}
T_{12} - T_1 &= \frac{T_1 T_2}{T_1 + T_2} - T_1 \\
&= \frac{T_1 T_2 - T_1^2 - T_1 T_2}{T_1 + T_2} \qquad\qquad (2.68) \\
&= -\frac{T_1^2}{T_1 + T_2} < 0,
\end{aligned}
$$

$$
\begin{aligned}
T_{123} - T_1 &= \frac{T_1 T_2 T_3}{T_1 T_2 + T_1 T_3 + T_2 T_3} - T_1 \\
&= \frac{T_1 T_2 T_3 - T_1^2 T_2 - T_1^2 T_3 - T_1 T_2 T_3}{T_1 T_2 + T_1 T_3 + T_2 T_3} \qquad\qquad (2.69) \\
&= \frac{-T_1^2 T_2 - T_1^2 T_3}{T_1 T_2 + T_1 T_3 + T_2 T_3} < 0.
\end{aligned}
$$

Inequalities for a comparison with pumps 2 and 3 ($i, j = 2$ or 3) are proved analogously. A comparison between scenarios with two and three pumps yields

$$T_{123} - T_{12} = \frac{T_1 T_2 T_3}{T_1 T_2 + T_1 T_3 + T_2 T_3} - \frac{T_1 T_2}{T_1 + T_2}$$

$$= \frac{T_1 T_2 T_3 (T_1 + T_2) - T_1 T_2 (T_1 T_2 + T_1 T_3 + T_2 T_3)}{T_1 T_2 + T_1 T_3 + T_2 T_3}$$

$$= \frac{T_1^2 T_2 T_3 + T_1 T_2^2 T_3 - T_1^2 T_2^2 - T_1^2 T_2 T_3 - T_1 T_2^2 T_3}{T_1 T_2 + T_1 T_3 + T_2 T_3}$$

$$= \frac{-T_1^2 T_2^2}{T_1 T_2 + T_1 T_3 + T_2 T_3} < 0. \tag{2.70}$$

We see that two pumps are more productive than one, and three are more productive than either two or one. This makes complete sense.

All four checks could be used to detect an error, should we have made one in solving these problems. For example, let us see what would happen if we obtained the following erroneous solution to the two-pump problem:

$$\tilde{T}_{12} = \frac{2 T_1 T_2}{T_1 + T_2}. \tag{2.71}$$

Below we go through all four checks and apply them to the erroneous solution above:

1. Suppose we have two pumps and pump 1 is many times more powerful than pump 2, which means that $T_1 \ll T_2$. Then equation (2.71) is reduced to $\tilde{T}_{12} = 2 T_1$, which is wrong: adding a weak second pump would not slow down draining the pool by pump 1.

2. Now we look at a case of three pumps when one of them (for example, pump 3) is very weak compared to the other two. As we have seen above, this case reduces to equation (2.66). It does not match equation (2.71), pointing to the existence of an error in one or both of these equations.

3. Next, we consider the problem with two pumps of equal power: $T_1 = T_2$. From equation (2.71), we get $\tilde{T}_{12} = T_1 = T_2$. This result is wrong, too: it predicts that two pumps working together would drain the pool in the same time as each of the pumps working separately.

4. Finally, we compute $\tilde{T}_{12} - T_1$ to get

$$\tilde{T}_{12} - T_1 = \frac{2 T_1 T_2}{T_1 + T_2} - T_1$$

$$= \frac{2 T_1 T_2 - T_1^2 - T_1 T_2}{T_1 + T_2} \tag{2.72}$$

$$= \frac{T_1 (T_2 - T_1)}{T_1 + T_2}.$$

This expression is positive if $T_2 > T_1$, which contradicts our intuition that two pumps should always be more productive than one.

These examples show how limiting cases point to the existence of an error. While they do not necessarily specify the exact nature of the error or ways to fix it, they are still very helpful. When we already know that there is an error, it usually becomes a matter of time and effort to locate and fix it.

2.11   The Sum of an Unknown and Its Reciprocal

Here we turn to the problem for the sum of an unknown and its reciprocal:

$$x + \frac{1}{x} = d, \tag{2.73}$$

which is solved in section A.8:

$$x_{1,2} = \frac{d \pm \sqrt{d^2 - 4}}{2}, \tag{2.74}$$

where subscripts $1, 2$ correspond to the $\pm$ signs in the right-hand side. There are several limiting and special cases to check:

1. What happens if d is large ($d \gg 1$)? By looking at equation (2.73), we see that a large value for d can be produced either if x is large ($x_1 \gg 1$) or if, conversely, x is small ($x_2 \ll 1$). Moreover, if $x_1 \gg 1$, we can neglect the $1/x_1$ term in the left-hand side of equation (2.73) and conclude that $x_1 \approx d$. Similarly, if x_2 is small, we should get $x_2 \approx 1/d$.

    The same must be true for the solution given by equation (2.74). If we set $d \gg 1$, then we can approximate $d^2 - 4 \approx d^2$, which does give us $x_1 \approx d$ for the plus sign in the solution. Presumably the minus sign should give us $x_2 \approx 1/d$. Rather than proving this directly from equation (2.74), which requires using calculus, we refer to section 3.7, where we shall see that from a symmetry property it follows that $x_1 x_2 = 1$. Then, if $x_1 \approx d$, we get $x_2 \approx 1/d$.

2. When does the problem have real-valued solutions? The criterion for this is that the discriminant $d^2 - 4$ must be nonnegative. This means that $d \geq 2$ or $d \leq -2$. Let us consider the case of positive d; the other case is analogous.

    If we substitute the limiting value $d = 2$ in equation (2.74), we will see that it yields $x_{1,2} = 1$. Let us now turn to the formulation of the problem. We have learned that it has no solutions for $d < 2$. Apparently, the sum of x and its reciprocal is always not smaller than 2. For $x \gg 1$, the sum is dominated by the first term and is correspondingly large; for $x \ll 1$ the sum is dominated by the second term and again has a large value. At $x = 1$, the sum $x + 1/x$ takes a minimum, which is equal to 2.

    This is an example of a limiting case that helps us clarify the behavior of the equation for different values of the variables in it. By using this limiting case, we defined the valid range of parameter d and obtained some insight in the behavior of the left-hand side of equation (2.73).

3. What happens if the right-hand side is expressed through some other parameter q in a special way: $d = q + 1/q$? In this case, the equation takes the following form:

$$x + \frac{1}{x} = q + \frac{1}{q}. \tag{2.75}$$

By examining this equation, we conclude that one of its roots must be equal to q. Let us now substitute $d = q + \frac{1}{q}$ in equation (2.74):

$$x_{1,2} = \frac{q + \frac{1}{q} \pm \sqrt{\left(q + \frac{1}{q}\right)^2 - 4}}{2}. \tag{2.76}$$

After expanding the square and collecting the terms, we get

$$x_{1,2} = \frac{q + \frac{1}{q} \pm \sqrt{\left(q - \frac{1}{q}\right)^2}}{2}. \tag{2.77}$$

This yields

$$\begin{aligned} x_1 &= q, \\ x_2 &= \frac{1}{q}. \end{aligned} \tag{2.78}$$

We do get one of the roots equal to q. Curiously, the other root happens to be equal to $1/q$. This is consistent with the symmetry implicit in the problem, which we will explore in section 3.7.

4. What happens if we square equation (2.73)? This analysis is prompted by observing a special property: squaring produces an equation that is in some respects similar to the original one:

$$\left(x + \frac{1}{x}\right)^2 = d^2. \tag{2.79}$$

After expanding the square and collecting the terms, we get

$$x^2 + \frac{1}{x^2} = d^2 - 2. \tag{2.80}$$

The left-hand side of this last equation is also a sum of an unknown x^2 and *its* reciprocal, and the right-hand side has a somewhat different form, but the general structure is similar. Therefore, its solution should be given by the same formula (2.74), where we should use $d^2 - 2$ in place of d and where we would get a solution for x^2 in place of x. We conclude that if we square solution (2.74) for x, we should get a result similar to it, but with $d^2 - 2$ in place of d. Let us check this prediction. Squaring equation (2.74) produces

$$
\begin{aligned}
x_{1,2}^2 &= \left(\frac{d \pm \sqrt{d^2 - 4}}{2} \right)^2 \\
&= \frac{d^2 \pm 2d \sqrt{d^2 - 4} + \left(\sqrt{d^2 - 4} \right)^2}{4} \\
&= \frac{2d^2 - 4 \pm 2 \sqrt{d^4 - 4d^2}}{4} \\
&= \frac{d^2 - 2 \pm \sqrt{d^4 - 4d^2 + 4 - 4}}{2} \\
&= \frac{\left(d^2 - 2 \right) \pm \sqrt{(d^2 - 2)^2 - 4}}{2}.
\end{aligned}
\tag{2.81}
$$

While it is not a miracle, it does feel as if a piece of a puzzle fell in place: replacing d with $d^2 - 2$ does produce a solution for x^2.

2.12   Designing Satellite Coverage

In this section we explore the problem for designing a satellite antenna to transmit a radio signal to terrestrial customers. This requires knowledge of the footprint of the signal on the ground.

Figure 2.7A shows a cross section of this problem. A satellite flies at altitude H above Earth and transmits a signal downward. The antenna beam has a conical shape, and the angle from the center to the edge of the beam is α. Earth's radius is R. The satellite beam covers a round spot on Earth's surface. Section A.24 presents a solution for the length L of the arc on Earth's surface from the center of the beam spot to its edge:

$$
L = R \left(\sin^{-1} \left(\frac{R + H}{R} \sin \alpha \right) - \alpha \right).
\tag{2.82}
$$

There are three parameters in this problem: R, H, and α. For practical purposes, Earth's radius R remains fixed, but an engineer can vary the orbit height H and the antenna half-beam width α. The last two parameters define the trade space here. In real life, there are additional considerations, such as the cost of launching a satellite to a particular orbit or having enough onboard power to "fill" a wide beam. Any comprehensive design should satisfy practical constraints while ensuring that the signal footprint is sufficient to serve the customers of that satellite broadcast.

The best way to achieve the desired coverage is to solve the problem in a general case first and then to analyze various options. Below we look at several limiting cases within

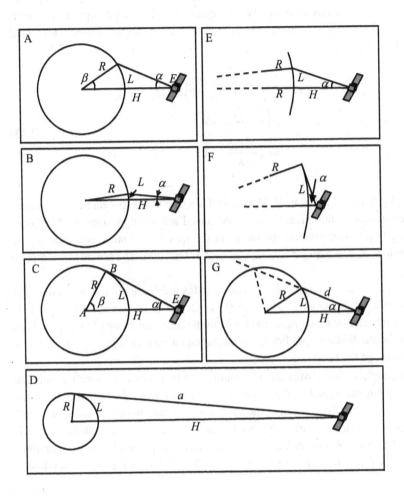

Figure 2.7
Designing satellite coverage: limiting cases

the trade space for orbit height H and half-beam width α. These limiting cases can become starting points to select the best option for satellite design.

The limiting cases that we consider for the satellite coverage problem are as follows:

1. We consider a very narrow satellite beam, $\alpha \ll 1$, while keeping the other parameters fixed (figure 2.7B). In this case we should expect a small beam footprint on Earth's surface. Specifically, we turn to equation (2.82) and use small angle approximations[4] for sin and for $\sin^{-1}$:

$$
\begin{aligned}
L &\approx R\left(\sin^{-1}\left(\frac{R+H}{R}\alpha\right) - \alpha\right) \\
&\approx R\left(\frac{R+H}{R}\alpha - \alpha\right) \\
&= R\alpha\left(\frac{R+H}{R} - 1\right) \\
&= \alpha H.
\end{aligned}
\tag{2.83}
$$

Since α is small, L is small as well. Moreover, for a tiny beam spot on Earth's surface, Earth's curvature does not affect the result because Earth is approximately "flat" over small areas. This would produce the radius of the spot $L \approx H \tan \alpha$. Since $\tan \alpha \approx \alpha$ for small values of α, we do get $L \approx \alpha H$, as above. This limiting case makes complete sense.

In practice, a narrow satellite beam concentrates the energy and delivers a higher signal intensity on the ground, which may be necessary for some customers. It is used by several satellite systems, including Iridium Communications.[5] Specific design parameters for the Iridium spot beam would use equation (2.83) or its more accurate version to manage the coverage.

2. What is the angle α that covers the maximum area but does not waste the onboard power? We want the signal to cover the full area on Earth's surface that faces the satellite, but we do not want much of the energy in the beam to pass beyond the globe. This design is shown in figure 2.7C, where the outer edge of the antenna beam is tangent to the globe. A line that is tangent to a circle is perpendicular to its radius. Hence, ABE is a right triangle, and $\beta = \pi/2 - \alpha$. The arc length L is then computed as follows:

$$
L = R\beta = R\left(\frac{\pi}{2} - \alpha\right).
\tag{2.84}
$$

4. Small angle approximations: if $x, y \ll 1$, then $\sin x \approx x$; $\sin^{-1} y \approx y$.

5. To be exact, each Iridium satellite produces multiple spot beams that jointly cover a larger area.

Comparing this equation to equation (2.82), we see that for this particular case we must have

$$\sin^{-1}\left(\frac{R+H}{R}\sin\alpha\right) = \frac{\pi}{2}. \tag{2.85}$$

This is the maximum possible value of the inverse sine, which requires that

$$\frac{R+H}{R}\sin\alpha = 1. \tag{2.86}$$

We conclude that the "max footprint, no waste" case corresponds to

$$\sin\alpha = \frac{R}{R+H}. \tag{2.87}$$

This is indeed the case for the right triangle *ABE* in figure 2.7C.

Note that we have no solutions for L if angle α is greater than the one specified by equation (2.87) because the argument of the inverse sine exceeds 1. Indeed, in this case, the outer edge of the signal beam has no intersections with Earth's surface. This limiting case is a close cousin of the one that we investigated in case 2 in section 2.5.

This design is used for many real satellites to ensure maximum coverage of the Earth's area, including GPS.[6]

3. Next, we consider the previous limiting case and place an additional condition: what is the "max footprint, no waste" scenario for a high-orbit satellite (see figure 2.7D)? This means that we add condition $H \gg R$ on top of the previous limiting case. Then equation (2.87) yields $\sin\alpha \ll 1$, or $\alpha \ll 1$. This is understandable: if a satellite is at a very large distance, we require a narrow angle to cover Earth. Note that a small value of $\sin\alpha$ in equation (2.85) is compensated by a large value of $(R+H)/R$, so that the value of the inverse sine is still equal to $\pi/2$. Therefore, in our general solution (2.82) we get

$$L = R\left(\sin^{-1}\left(\frac{R+H}{R}\sin\alpha\right) - \alpha\right) \approx R\frac{\pi}{2}. \tag{2.88}$$

This means that a satellite at a large altitude can cover almost half of the globe. That is the case for many geostationary satellites, which fly at high altitudes and are widely used for telecommunications and TV broadcast.

4. Next, we look at the case of a low-Earth orbit satellite: $H \ll R$ (figure 2.7E). We can see that for a satellite at a low orbit, it is impossible to get a large beam footprint on

---

6. The actual design of the GPS beam is a bit more complex. The beam boundary is never sharp; a real-life beam has a gradual roll-off over some range of angles, but engineers have made a good effort to direct the GPS signal to the widest possible area on Earth without spilling much of it over to the free space.

Earth's surface, and we get a small value for L. If we now look at equation (2.82) and use condition $H \ll R$, we will see that

$$\frac{R+H}{R} \approx 1. \tag{2.89}$$

This leads to

$$\sin^{-1}\left(\frac{R+H}{R}\sin\alpha\right) - \alpha \approx \sin^{-1}(\sin\alpha) - \alpha = 0. \tag{2.90}$$

The expression on the left is therefore approximately equal to zero, which means that its absolute value is small. Using this last condition in equation (2.82) shows that we do get a small[7] footprint: $L \ll R$.

This creates a problem for the designer. A link from a satellite to a ground terminal is possible only if this terminal is within the beam coverage, and the coverage is severely limited in the case of a low-Earth orbit. For example, when such a satellite flies over the ocean, it cannot connect to the command center, simply because there is no receiving equipment within the beam below the satellite. To ensure uninterruptible connection to such a satellite, a designer must use other means, such as using separate equipment to connect this spacecraft to another satellite or satellites in the direct view. Both the Iridium system and the *International Space Station* (ISS) use this approach. The Iridium system relays the signal from one Iridium satellite to another, and the ISS connects to a geostationary satellite at a much higher altitude.

5. Next, we combine limiting cases 4 and 2. What happens for a low-Earth orbit satellite ($H \ll R$) if we want to cover the maximum area? We look at equation (2.85), which is transformed as follows:

$$\frac{\pi}{2} = \sin^{-1}\left(\frac{R+H}{R}\sin\alpha\right) \approx \sin^{-1}(\sin\alpha) = \alpha. \tag{2.91}$$

We see that the half-angle of the antenna beam width is close to $\pi/2$, which means that the satellite beam spans nearly a hemisphere. Figure 2.7F illustrates this result. The designer has to select an antenna that can create such a wide beam.

The length of the arc on Earth's surface from the center of the beam spot to its edge L can be approximately computed by looking at figure 2.7C if we assume that H is small there (that is, $H \ll R$). In that case, arc L is approximately equal to line segment BE. Since ABE is a right triangle, we can find BE from the Pythagorean theorem:

$$BE^2 = AE^2 - AB^2. \tag{2.92}$$

Since $AB = R$ and $AE = R + H$, we get

---

7. Note that the satellite footprint is small only as compared to Earth's radius, but its numerical value in meters can appear to be large.

$$BE = \sqrt{(R + H)^2 - R^2} = \sqrt{2RH + H^2}. \tag{2.93}$$

For a low-orbit satellite, $H \ll R$ and therefore $H^2 \ll 2RH$. We also recall that $L \approx BE$. This yields

$$L \approx \sqrt{2RH}. \tag{2.94}$$

This design is adopted by the ORBCOMM satellite constellation. Multiple satellites in the constellation are placed on orbits so that their combined coverage blankets Earth's surface. Equation (2.94) would be handy in estimating the number of satellites that are required for global coverage and in designing their placement.

6. Finally, we explore a spurious solution to this problem. Here we refer to equation (A.109) in section A.24:

$$\sin(\beta + \alpha) = \frac{R + H}{R} \sin \alpha. \tag{2.95}$$

Technically, it has more solutions for β than the one we used to obtain equation (2.82). We selected the solution in hand because we know that β must lie between $-\pi/2$ and $+\pi/2$. In addition to that one, there is a solution $\beta' + \alpha = \pi - (\beta + \alpha)$, and all values are obtained from these two values of β by adding $2\pi n$, where n is an integer. Is there any significance to these solutions?

For the angles that are obtained by adding $2\pi n$ to the original solution, we get additional full revolutions that do not change the geometry of the problem. However, the solution given by $\beta' + \alpha = \pi - (\beta + \alpha)$ is a bit more subtle. Looking at figure 2.7G, we see that it corresponds to the *second* intersection of the outer edge of the beam with Earth's surface. In practice, this area cannot be served by the satellite because it is obscured by Earth, even though this solution formally satisfies equation (2.95). However, the equations "do not know" that a radio signal cannot penetrate through Earth. In a hypothetical example of a satellite that emits neutrinos in a cone beam, the second solution would be perfectly valid because neutrinos are able to go through Earth with a low loss. Of course, this special case does not have a practical implementation, yet it is helpful for understanding the solution better.

This completes our exploration of limiting cases for satellite coverage equations. As you can see, limiting cases provide a wealth of information on the problem. By looking at it from different angles (pun intended), we get a better understanding of various possibilities here. Many of these theoretical possibilities found their way into real satellite systems.

2.13   Two Circles Inscribed in an Angle

Consider two circles that are inscribed in an angle in such a way that they touch each other (figure 2.8A). What is the ratio R/r of their radii as a function of the angle? Section A.26 presents the following solution of this problem:

$$\frac{R}{r} = \frac{1 + \sin\frac{\beta}{2}}{1 - \sin\frac{\beta}{2}}. \tag{2.96}$$

In this section, we consider two limiting cases:

1. Small angle limiting case: $\beta \to 0$. This case is illustrated in figure 2.8B. From the figure, we expect the ratio of the radii to be close to 1. Indeed, if we apply condition $\beta \to 0$ to equation (2.96), we observe that $\sin(\beta/2) \to 0$, and $R/r \to 1$ as a result.

2. Large angle limiting case: $\beta \to \pi$. This case is illustrated in figure 2.8C. In contrast with the previous limiting case, we expect the ratio of the radii to be large. The condition $\beta \to \pi$ in equation (2.96) produces $\sin(\beta/2) \to \sin(\pi/2) = 1$, and $R/r \to \infty$.

We see that both limiting cases hold.

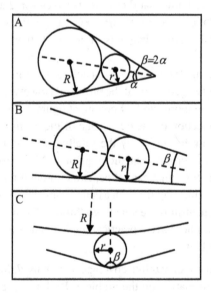

Figure 2.8
Two circles inscribed in an angle: limiting cases

2.14   The Intersections between a Circle and a Parabola

Here we consider the problem of finding intersections between a circle and a parabola (see also section A.6). Contrary to our usual practice to provide just a formulation, a final result, and a reference to a solution in the appendix, this section presents what appears to be a full derivation. This derivation has an error deliberately introduced (section A.6 does contain a correct solution), to show how an error can be detected by using a limiting case.

Key Point

Limiting cases are a powerful means to find errors in your solutions.

A circle and a parabola are defined by the following equations (see figure 2.9):

$$x^2 + y^2 = R^2,$$
$$y = gx^2 + y_0. \tag{2.97}$$

We are seeking the coordinates of the intersections of these two curves (if any). To solve for x, y, we substitute the x^2 from the first equation in the second one to get a quadratic equation for y:

$$y = g(R^2 - y^2) + y_0. \tag{2.98}$$

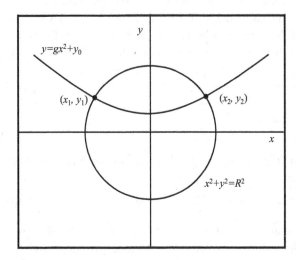

Figure 2.9
A circle and a parabola: limiting cases

We rearrange the terms to get

$$gy^2 + y - gR^2 - y_0 = 0. \tag{2.99}$$

For a nonnegative discriminant, the real-valued solutions are given by the quadratic formula:

$$y_{1,2} = \frac{-1 \pm \sqrt{1 + g(gR^2 + y_0)}}{2g}, \tag{2.100}$$

where subscripts $1, 2$ correspond to the $\pm$ signs in the right-hand side. We substitute these solutions in the first equation of (2.97) to get

$$x_{1,2} = +\sqrt{R^2 - y_{1,2}^2},$$
$$x_{3,4} = -\sqrt{R^2 - y_{1,2}^2}. \tag{2.101}$$

In addition to the previous requirement of having a nonnegative discriminant, note that the solutions are valid only for such values of y that satisfy condition $R^2 \geq y_{1,2}^2$ or, equivalently, $R \geq |y_{1,2}|$. Therefore, if the discriminant $1 + g(gR^2 + y_0)$ is negative, equations (2.97) have no solutions; if the discriminant is nonnegative and only one of the values of $y_{1,2}$ satisfies condition $R \geq |y_{1,2}|$, equations (2.97) have two solutions; finally, if both values of $y_{1,2}$ satisfy condition $R \geq |y_{1,2}|$, equations (2.97) have four solutions.

Let us now consider a limiting case. In the equation for the parabola, y_0 is the offset from the origin. If the offset is equal to the circle radius R, we should see a case that is shown in figure 2.10, and at least one of the roots must be $x = 0, y = R$. Therefore, for $y_0 = R$, equation (2.100) should have the following form for at least one choice of the sign:

$$\frac{-1 \pm \sqrt{1 + g(gR^2 + R)}}{2g} = R. \tag{2.102}$$

We transform this to

$$\pm \sqrt{1 + g(gR^2 + R)} = 2gR + 1. \tag{2.103}$$

Squaring this equation and expanding the left-hand side produces the following result, which should be valid for any value of g:

$$g^2R^2 + gR + 1 = 4g^2R^2 + 4gR + 1 \quad \text{✍}. \tag{2.104}$$

However, if $g, R \neq 0$, this last equation cannot be satisfied. We conclude that there must be an error in equation (2.100). Moreover, we see that the left-hand side of equation (2.104) lacks a multiplier of 4 for two terms. This gives us a clue about the nature of the error that we have made in the derivation. Indeed, when we applied the quadratic formula to produce equation (2.100), we missed the factor of 4. The correct solution is as follows (it is also given in section A.6):

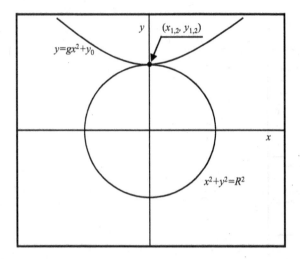

Figure 2.10
Detecting an error in the circle and parabola problem

$$y_{1,2} = \frac{-1 \pm \sqrt{1 + 4g(gR^2 + y_0)}}{2g}. \tag{2.105}$$

A check of the limiting case $y_0 = R$ shows that condition $y_{1,2} = R$ is now satisfied for one of the roots. As you can see, checking a limiting case was instrumental in detecting this error.

2.15 Linear Regression

A linear regression is a popular algorithm to interpret numerical data. Section A.33 presents a basic version of this algorithm that deals with N measurements for variables x and y. These variables are not independent, and we assume that they have an underlying linear relationship. Because of measurement errors or other factors, the points do not lie exactly on a straight line but may deviate somewhat from it (see figure 2.11A). Mathematically, the link between x and y is described by

$$y = ax + b + R, \tag{2.106}$$

where R is a random variable.[8]

---

8. See section 7.1 for an explanation of a random variable.

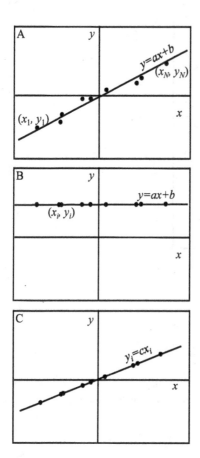

Figure 2.11
Linear regression: limiting cases

The linear regression algorithm estimates the best fit for model parameters a and b from available data:

$$a = \frac{N \sum_{i=1}^{N} x_i y_i - \sum_{i=1}^{N} x_i \cdot \sum_{i=1}^{N} y_i}{N \sum_{i=1}^{N} x_i^2 - \left(\sum_{i=1}^{N} x_i\right)^2},$$

$$b = \frac{\sum_{i=1}^{N} y_i \cdot \sum_{i=1}^{N} x_i^2 - \sum_{i=1}^{N} x_i \cdot \sum_{i=1}^{N} x_i y_i}{N \sum_{i=1}^{N} x_i^2 - \left(\sum_{i=1}^{N} x_i\right)^2},$$

(2.107)

where x_i, y_i are the available data for variables x and y.

Let us check if these formulas make intuitive sense. We look at two limiting cases:

1. We see what happens if all y_i have the same value, which we denote as $\bar{y}$. On the plot, all data points would lie on a horizontal line (figure 2.11B). We expect the best-fit line to coincide with that line, which would mean that $a = 0$ and $b = \bar{y}$. Do we get these values from equations (2.107)? Setting $y_i = \bar{y}$ for all values of i yields

$$
a = \frac{N \sum_{i=1}^{N} x_i \bar{y} - \sum_{i=1}^{N} x_i \cdot \sum_{i=1}^{N} \bar{y}}{N \sum_{i=1}^{N} x_i^2 - \left(\sum_{i=1}^{N} x_i \right)^2},
$$
$$
b = \frac{\sum_{i=1}^{N} \bar{y} \cdot \sum_{i=1}^{N} x_i^2 - \sum_{i=1}^{N} x_i \cdot \sum_{i=1}^{N} x_i \bar{y}}{N \sum_{i=1}^{N} x_i^2 - \left(\sum_{i=1}^{N} x_i \right)^2}.
$$
(2.108)

We observe that $\sum_{i=1}^{N} \bar{y} = N \bar{y}$. We also factor out $\bar{y}$ from other sums to get

$$
a = \frac{N \bar{y} \sum_{i=1}^{N} x_i - N \bar{y} \sum_{i=1}^{N} x_i}{N \sum_{i=1}^{N} x_i^2 - \left(\sum_{i=1}^{N} x_i \right)^2} = 0
$$
(2.109)

and

$$
b = \frac{N \bar{y} \sum_{i=1}^{N} x_i^2 - \bar{y} \sum_{i=1}^{N} x_i \cdot \sum_{i=1}^{N} x_i}{N \sum_{i=1}^{N} x_i^2 - \left(\sum_{i=1}^{N} x_i \right)^2}
$$
$$
= \bar{y} \frac{N \sum_{i=1}^{N} x_i^2 - \left(\sum_{i=1}^{N} x_i \right)^2}{N \sum_{i=1}^{N} x_i^2 - \left(\sum_{i=1}^{N} x_i \right)^2} = \bar{y}.
$$
(2.110)

This limiting case works as expected.

2. Now let us see what happens if the data fall exactly on a slant line (figure 2.11C):

$$
y_i = c x_i.
$$
(2.111)

This corresponds to $b = 0$ and to having no random term in equation (2.106). In this case, we expect the best-fit line to exactly reproduce that linear relationship, yielding $a = c, b = 0$. Let us see if this follows from the linear regression equations. We substitute $y_i = c x_i$ in equations (2.107) to get

$$
a = \frac{N \sum_{i=1}^{N} x_i c x_i - \sum_{i=1}^{N} x_i \cdot \sum_{i=1}^{N} c x_i}{N \sum_{i=1}^{N} x_i^2 - \left(\sum_{i=1}^{N} x_i \right)^2} = c \frac{N \sum_{i=1}^{N} x_i^2 - \left(\sum_{i=1}^{N} x_i \right)^2}{N \sum_{i=1}^{N} x_i^2 - \left(\sum_{i=1}^{N} x_i \right)^2} = c \quad (2.112)
$$

and

$$b = c \frac{\sum_{i=1}^{N} x_i \cdot \sum_{i=1}^{N} x_i^2 - \sum_{i=1}^{N} x_i \cdot \sum_{i=1}^{N} x_i x_i}{N \sum_{i=1}^{N} x_i^2 - \left(\sum_{i=1}^{N} x_i\right)^2} = 0. \tag{2.113}$$

This limiting case also holds.

When using the linear regression algorithm, we want the results to "make sense" in addition to satisfying formal mathematical conditions. The two limiting cases above show that this algorithm works in an intuitive, reasonable way.

2.16   Summary

A limiting case analysis exploits an extreme or a special value of a variable. Often the problem is simplified at a limiting case. By checking a limiting case, we achieve multiple goals:

1. We can check the solution for errors. If we establish a limiting case for the initial formulation of the problem, then we can check it for any intermediate step of the solution and for the final result.

2. A limiting case analysis provides an economical means of playing out "what if" scenarios. By looking at different options, we gain a better understanding of the problem in hand. As I noted in the preface, real-life problems rarely result in a single-number solution. Usually the final result is a combination of trade-offs. To design a receiver in a cell phone, an engineer must consider its operation near a cell tower and at large distances from it, in an urban setting and in the woods, and so on. Limiting cases offer an opportunity to quickly scan our options and to predict the outcomes.

3. Limiting cases often define the domain of applicability for variables in the problem. Keeping track of where the solution is applicable is extremely important: it guards us against using it in scenarios where it does not work.

4. A limiting case can be a foothold for solving a tough problem. When you do not know where to start, define a limiting case that simplifies your problem and start from there.

Exercises

Problems with an asterisk present an extra challenge.

1. Use limiting cases $a \to b$ and $a \to -b$ to select the two correct formulas from the four options below. (Hint: For each equation, select a limiting case that nulls its left-hand side.)

   a) $a^3 - b^3 = (a + b)(a^2 + ab + b^2)$

   b) $a^3 - b^3 = (a - b)(a^2 + ab + b^2)$

   c) $a^3 + b^3 = (a + b)(a^2 - ab + b^2)$

   d) $a^3 + b^3 = (a - b)(a^2 - ab + b^2)$

2. Check a limiting case $\alpha = 0$ for the following trigonometric identities. (Hint: Use $\sin(0) = 0$ and $\cos(0) = 1$.)

   a) $\sin(\alpha + \beta) = \sin\alpha\cos\beta + \sin\beta\cos\alpha$

   b) $\sin(\alpha - \beta) = \sin\alpha\cos\beta - \sin\beta\cos\alpha$

   c) $\cos(\alpha + \beta) = \cos\alpha\cos\beta - \sin\alpha\sin\beta$

   d) $\cos(\alpha - \beta) = \cos\alpha\cos\beta + \sin\alpha\sin\beta$

3. Check a limiting case $\alpha = 0$ for the following trigonometric identities. (Hint: Use $\tan(0) = 0$ and $\cot(\alpha) \to \infty$ when $\alpha \to 0$.)

   a) $\tan(\alpha + \beta) = \dfrac{\tan\alpha + \tan\beta}{1 - \tan\alpha\tan\beta}$

   b) $\tan(\alpha - \beta) = \dfrac{\tan\alpha - \tan\beta}{1 + \tan\alpha\tan\beta}$

   c) $\cot(\alpha + \beta) = \dfrac{\cot\alpha\cot\beta - 1}{\cot\beta + \cot\alpha}$

   d) $\cot(\alpha - \beta) = \dfrac{\cot\alpha\cot\beta + 1}{\cot\beta - \cot\alpha}$

4. Check limiting cases for $\alpha \to 0$ and $\alpha \to \pi$ for the following trigonometric identities:

   a) $\sin 2\alpha = 2\cos\alpha\sin\alpha$

   b) $\cos 2\alpha = \cos^2\alpha - \sin^2\alpha$

   c) $\sin^2\dfrac{\alpha}{2} = \dfrac{1 - \cos\alpha}{2}$

   d) $\cos^2\dfrac{\alpha}{2} = \dfrac{1 + \cos\alpha}{2}$

5. Here we revisit the problems for the sum of two ratios and the sum of two scaled ratios. They are solved in sections A.10 and A.11 and compared in section 2.7. Check that each equation in

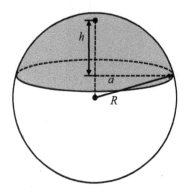

Figure 2.12
A spherical cap: limiting cases

the first column of table 2.1 in section 2.7 can be obtained from the corresponding equation in the second column in the limiting case $p \to 1, q \to 1$.

6. A spherical cap is the part of a sphere that lies above a plane that crosses this sphere (see figure 2.12 and section A.30). The volume V and the surface area S of a spherical cap are given by

$$V = \frac{1}{3}\pi h^2(3R - h),$$
$$S = 2\pi Rh. \tag{2.114}$$

Explore limiting cases $h \to 0, h = R$, and $h \to 2R$. Corresponding formulas for a sphere are

$$V_{\text{sphere}} = \frac{4}{3}\pi R^3,$$
$$S_{\text{sphere}} = 4\pi R^2. \tag{2.115}$$

7. The law of cosines computes the length c of a side of a triangle using lengths a, b of two other sides and the angle γ between these two sides (figure 2.13):

$$c^2 = a^2 + b^2 - 2ab\cos\gamma. \tag{2.116}$$

Check the following limiting cases and provide an interpretation for each:

a) $a \to 0$

b) $a \to \infty$ and c remains finite

c) $\gamma \to 0$

d) $\gamma \to \frac{\pi}{2}$

e) $\gamma \to \pi$

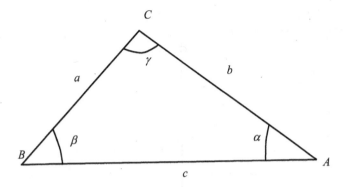

Figure 2.13
Law of cosines: limiting cases

8. Section A.12 solves the following equation:

$$\frac{1}{x-a} - \frac{1}{x-b} = d. \tag{2.117}$$

The solution is as follows:

$$x_{1,2} = \frac{d(a+b) \pm \sqrt{d^2(a-b)^2 + 4d(a-b)}}{2d}, \tag{2.118}$$

where subscripts $1,2$ correspond to the $\pm$ signs in the right-hand side. Formulate and check limiting cases for equation (2.117) and its solution (2.118). How many limiting cases can you come up with? (Hint: Use the analysis in section 2.6 as a template.)

9. A radar measures range (distance) to the object it is tracking. Assume that there are two radars that detect a sea vessel at ranges R_1 and R_2 (see figure 2.14). The respective coordinates of the radars are $x_1 = 0, y_1 = 0$ and $x_2 = D, y_2 = 0$. Section A.25 provides a solution for the coordinates of the vessel as

$$x = \frac{D^2 + R_1^2 - R_2^2}{2D},$$
$$y = \pm\sqrt{R_1^2 - x^2}. \tag{2.119}$$

Explore the effect of the following limiting cases on x and y, and explain the meaning of each:

a) $R_2^2 = D^2 + R_1^2$

b) $R_1^2 = D^2 + R_2^2$

c) $D = R_1 + R_2$

d) $R_1 = R_2$

e) $R_1 = D + R_2$

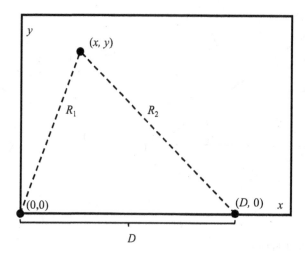

Figure 2.14
Detecting a vessel by two radars: limiting cases

    f) $R_2 = D + R_1$

10. A torus is a donut-shaped body (see figure 2.15). Use limiting cases to flag the incorrect formulas for the volume of a torus. Assume that $R > r$.

    a) $V = 2\pi^2 \left(R^3 + r^3 \right)$

    b) $V = 2\pi^2 R r^2$

    c) $V = 2\pi r^3 e^{-\frac{R}{r}}$

    d) $V = 2\pi R^3 e^{-\frac{r}{R}}$

11. Section A.15 solves an equation for the ratio of two cosine functions:

$$\frac{\cos(\alpha + x)}{\cos(\alpha - x)} = \frac{p}{q}. \tag{2.120}$$

The solution is given by

$$x = \tan^{-1}\left(\cot\alpha \cdot \frac{q - p}{q + p} \right) + n\pi, \tag{2.121}$$

where n is an integer.

    a)   Check the limiting case $p \to q$ for both the formulation of the problem and the solution.

    b)   Note that the cosine function is periodic with the period of 2π, which implies that for any solution x, the value $x + 2\pi m$ is also a solution. Yet, equation (2.121) has a term $n\pi$, where n

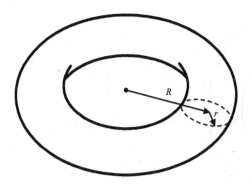

Figure 2.15
A torus: limiting cases

may be an odd number. Would odd values of n make the equation not valid if you substitute x in the limiting case $p \to q$ from equation (2.121) in equation (2.120)?

c) Use limiting cases to flag the wrong solutions among the following options and explain how each limiting case works:

i. $x = \tan^{-1}\left(\cot\alpha \cdot \dfrac{q-p}{p+q}\right) + n\pi$

ii. $x = \tan^{-1}\left(\cot\alpha \cdot \dfrac{p+q}{p-q}\right) + n\pi$

iii. $x = \cos^{-1}\left(\sin\alpha \cdot \dfrac{q-p}{p+q}\right) + n\pi$

iv. $x = \sin^{-1}\left(\cos\alpha \cdot \dfrac{q-p}{p+q}\right) + n\pi$

v. $x = \cos^{-1}\left(\sin\alpha \cdot \dfrac{q+p}{p-q}\right) + n\pi$

vi. $x = \sin^{-1}\left(\cos\alpha \cdot \dfrac{q+p}{p-q}\right) + n\pi$

12*. In section A.14 we seek a solution for an equation that contains a sum of two sine functions:

$$p\sin(x+\phi) + q\sin x = c. \qquad (2.122)$$

The solution is given as

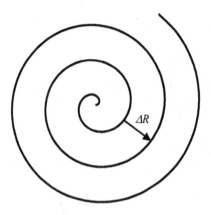

Figure 2.16
Archimedes's spiral: limiting cases

$$x = -\tan^{-1}\frac{p\sin\phi}{p\cos\phi + q} + (-1)^n \sin^{-1}\left(\frac{c}{\sqrt{p^2 + 2pq\cos\phi + q^2}}\right) + n\pi. \qquad (2.123)$$

a) Check the validity of the solution using the following limiting cases:

    i. $\phi \to 0$

    ii. $q \to 0$

    iii. $p \to 0$

b) Use a limiting case $\phi = -\pi/2$. Show that this problem is reduced to one that is solved in section A.13. Use the solution in section A.13 to check this limiting case for equation (2.123).

13. The radius of Archimedes's spiral increases linearly with the turn angle (figure 2.16). Use limiting cases to flag the incorrect formulas for the arc length of the spiral:

    a) $L = \dfrac{1}{4\pi}(\Delta R + 1)\left(\theta\sqrt{1 + \theta^2} + \ln\left(\theta + \sqrt{1 + \theta^2}\right)\right)$

    b) $L = \dfrac{1}{4\pi}\Delta R\left(\theta\sqrt{1 + \theta^2} + \ln\left(\theta + \sqrt{1 + \theta^2}\right)\right)$

c) $L = \dfrac{1}{4\pi}\Delta R\left(\theta\sqrt{1+\theta^2} + \ln\left(\theta\right)\right)$

d) $L = \dfrac{1}{4\pi}\Delta R\left(\dfrac{1+\theta^2}{\theta} + \ln\left(\theta + \sqrt{1+\theta^2}\right)\right)$

Here ΔR is the distance between the adjacent loops and θ is the total turn angle.

14. Chapter 3 shows how to solve the following trigonometric equation:

$$\sin(x-a) + \sin(x-b) = c. \qquad (2.124)$$

Use a limiting case $a \to b$ to select the correct solution among the following options:

a) $x = \dfrac{a+b}{2} + (-1)^n \sin^{-1}\left(\dfrac{c}{2\cos\frac{a-b}{2}}\right) + \pi n$

b) $x = \dfrac{a-b}{2} + (-1)^n \sin^{-1}\left(\dfrac{c}{2\cos\frac{a+b}{2}}\right) + \pi n$

c) $x = \dfrac{a-b}{2} + (-1)^n \sin^{-1}\left(\dfrac{c}{2\cos\frac{a-b}{2}}\right) + \pi n$

d) $x = \dfrac{a+b}{2} + (-1)^n \sin^{-1}\left(\dfrac{c}{2\cos\frac{a+b}{2}}\right) + \pi n$

15. There are syrups with masses $m_1, m_2,$ and m_3 with sugar concentrations $p_1, p_2,$ and p_3. Sections A.16 and A.17 show that blends of two or three of these syrups will have sugar concentrations respectively given by

$$p_{12} = \frac{p_1 m_1 + p_2 m_2}{m_1 + m_2},$$
$$p_{123} = \frac{p_1 m_1 + p_2 m_2 + p_3 m_3}{m_1 + m_2 + m_3}. \qquad (2.125)$$

a) Investigate the following limiting cases for blending two syrups and explain the results:

    i. The amount of syrup 2 is very small ($m_2 \ll m_1$).

    ii. Both syrups have the same concentration of sugar ($p_1 = p_2$).

    iii. Show that the sugar concentration of the mix is always in between the concentrations of the mixing ingredients ($p_1 \le p_{12} \le p_2$ or $p_1 \ge p_{12} \ge p_2$). (Hint: Try to compute $(p_{12} - p_1)(p_{12} - p_2)$ and show that the result is nonpositive.)

b) Show that if the amount of the third syrup is small ($m_3 \ll m_1; m_3 \ll m_2$), the equation for the problem for three syrups reduces to that for the problem for two syrups. Explain this result.

16. Section A.4 solves the problem of finding intersections between a circle and an ellipse (see figure 2.17). The circle and the ellipse are given by the following equations:

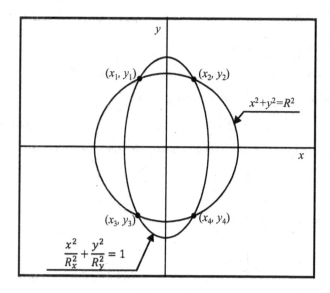

Figure 2.17
A circle and an ellipse: limiting cases

$$x^2 + y^2 = R^2,$$
$$\frac{x^2}{R_x^2} + \frac{y^2}{R_y^2} = 1. \tag{2.126}$$

Formulate one or more limiting cases for this problem and use them to flag the wrong solutions among the following options:

a) $x = \pm \sqrt{R_y^2 \dfrac{R_x^2 - R^2}{R_x^2 - R_y^2}}$; $\quad y = \pm \sqrt{R_x^2 \dfrac{R_y^2 - R^2}{R_y^2 - R_x^2}}$

b) $x = \pm \sqrt{R_x^2 \dfrac{R_y^2 - R^2}{R_y^2 - R_x^2}}$; $\quad y = \pm \sqrt{R_y^2 \dfrac{R_x^2 - R^2}{R_x^2 - R_y^2}}$

c) $x = \pm \sqrt{R_x^2 \dfrac{R_y^2 + R^2}{R_y^2 - R_x^2}}$; $\quad y = \pm \sqrt{R_y^2 \dfrac{R_x^2 + R^2}{R_x^2 - R_y^2}}$

(Hint: Note that the top and the bottom points on the ellipse are given by $y_{t,b} = \pm R_y$, and the rightmost and leftmost points are given by $x_{r,l} = \pm R_x$.)

17. We again refer to section A.4 and figure 2.17. The circle and the ellipse are given by the following equations:

$$x^2 + y^2 = R^2,$$
$$\frac{x^2}{R_x^2} + \frac{y^2}{R_y^2} = 1. \tag{2.127}$$

The coordinates of the intersection points (if any) for these two curves are given by

$$x = \pm \sqrt{R_x^2 \frac{R_y^2 - R^2}{R_y^2 - R_x^2}},$$
$$y = \pm \sqrt{R_y^2 \frac{R_x^2 - R^2}{R_x^2 - R_y^2}}. \tag{2.128}$$

Show that solutions exist if one of the following conditions holds:

$$R_x \leq R \leq R_y,$$
$$R_y \leq R \leq R_x. \tag{2.129}$$

Explain why this is true from figure 2.17. (Hint: Note that the top and the bottom points on the ellipse are given by $y_{t,b} = \pm R_y$, and the rightmost and leftmost points are given by $x_{r,l} = \pm R_x$.)

18*. Consider the problem for the difference between an unknown and its reciprocal:

$$x - \frac{1}{x} = d. \tag{2.130}$$

Its solution is presented in section A.9:

$$x_{1,2} = \frac{d \pm \sqrt{d^2 + 4}}{2}. \tag{2.131}$$

Following the examples in section 2.11, explore the following limiting cases in this problem:

a) A large value in the right-hand side: $d \to \infty$

b) A special case when d can be represented as $d = q - 1/q$, where q is a parameter

c) What happens if we square equation (2.130)? Show that the roots for the squared equation form two pairs and are reciprocal in each pair. Prove the reciprocity of x_1^2 and x_2^2 from equation (2.131).

19. Section A.5 solves the problem of finding the intersections between a circle and a hyperbola (see figure 2.18). The circle and the hyperbola are given by the following equations:

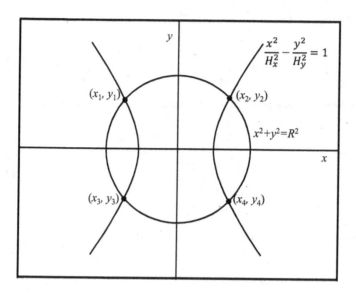

Figure 2.18
A circle and a hyperbola: limiting cases

$$x^2 + y^2 = R^2,$$
$$\frac{x^2}{H_x^2} - \frac{y^2}{H_y^2} = 1. \tag{2.132}$$

a) What is the value of x that satisfies the hyperbola equation for the limiting case $y = 0$?

b) Use the answer to the previous question to identify the wrong solutions for the intersections between the circle and the hyperbola:

i. $y = \pm \sqrt{H_y^2 \dfrac{R^2 - H_x^2}{H_x^2 + H_y^2}};$ $x = \pm \sqrt{H_x^2 \dfrac{R^2 - H_y^2}{H_x^2 + H_y^2}}$

ii. $y = \pm \sqrt{H_y^2 \dfrac{R^2 + H_x^2}{H_x^2 + H_y^2}};$ $x = \pm \sqrt{H_x^2 \dfrac{R^2 + H_y^2}{H_x^2 + H_y^2}}$

iii. $y = \pm \sqrt{H_y^2 \dfrac{R^2 - H_x^2}{H_x^2 + H_y^2}};$ $x = \pm \sqrt{H_x^2 \dfrac{R^2 + H_y^2}{H_x^2 + H_y^2}}$

iv. $y = \pm \sqrt{H_y^2 \dfrac{R^2 + H_x^2}{H_x^2 + H_y^2}};$ $x = \pm \sqrt{H_x^2 \dfrac{R^2 - H_y^2}{H_x^2 + H_y^2}}$

20. The law of sines links the angles and lengths of adjacent sides in a triangle (figure 2.13):

$$a \sin \beta = b \sin \alpha. \tag{2.133}$$

Check the following limiting cases and provide an interpretation for each:

a) $a \to 0$

b) $\alpha \to \dfrac{\pi}{2}$

c) $\alpha \to 0$

d) $\alpha \to \beta$

e) $\alpha \to \dfrac{\pi}{2} - \beta$

21. The law of tangents is another way to link the angles and lengths of adjacent sides in a triangle (figure 2.13):

$$\frac{a-b}{a+b} = \frac{\tan \frac{\alpha-\beta}{2}}{\tan \frac{\alpha+\beta}{2}}. \tag{2.134}$$

Check the following limiting cases and provide an interpretation for each:

a) $\beta \to 0$ and α remains finite

b) $a = b$

c) $\alpha \to \pi/2 - \beta$ (Hint: You may need to use a formula for the tangent of the difference of two angles from exercise 3.)

22*. The law of tangents above is closely related to Mollweide's formula (see figure 2.13 for notations):

$$\frac{a+b}{c} = \frac{\cos \frac{\alpha-\beta}{2}}{\sin \frac{\gamma}{2}},$$
$$\frac{a-b}{c} = \frac{\sin \frac{\alpha-\beta}{2}}{\cos \frac{\gamma}{2}}. \tag{2.135}$$

Explore the following limiting cases:

a) $c \to 0$, and a, b remain finite.

b) $a \to b$ (two sides are equal).

c) For the first Mollweide's formula, $c \to a + b$ and a, b, c remain finite.

d) For the second Mollweide's formula, $c \to a - b$ and a, b, c remain finite.

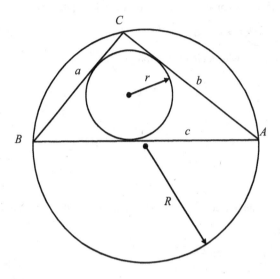

Figure 2.19
A triangle, an inscribed circle, and a circumscribed circle: limiting cases

23*. Heron's formula links the area S of a triangle, its half-perimeter $p = (a + b + c)/2$, and the lengths of its sides (see figure 2.13):

$$S = \sqrt{p(p - a)(p - b)(p - c)}. \tag{2.136}$$

Explore the following limiting cases:

a) $a \rightarrow 0, \quad b \rightarrow c$

b) $a \rightarrow b + c$

c) $a^2 + b^2 = c^2$

d) $a^2 + c^2 = b^2$

24. It is known that the radius of the inscribed circle r is as follows (see figure 2.19):

$$r = \frac{S}{p}, \tag{2.137}$$

where S is the area of the triangle and p is its half-perimeter: $p = (a + b + c)/2$. Explore the limiting case $S \rightarrow 0$ and p remains finite.

25. Another formula for the radius of the inscribed circle r is given by

$$r = \frac{\sqrt{p(p-a)(p-b)(p-c)}}{p}, \tag{2.138}$$

where p is its half-perimeter: $p = (a+b+c)/2$. Explore what happens if $p \to c$. Show that r becomes small in this case and why that follows from figure 2.19.

26. We again refer to the drawing in figure 2.19. The radius of the circumscribed circle R is given by

$$R = \frac{abc}{\sqrt{(a+b+c)(-a+b+c)(a-b+c)(a+b-c)}}. \tag{2.139}$$

Consider the following limiting cases:

a) Three separate limiting cases that lead to a small value in the denominator even if lengths of all sides remain finite:

   i. $a \to b + c$

   ii. $a \to b - c$

   iii. $a \to c - b$

   Why does the radius of the circumscribed circle become large in all these cases? Show this from the equation for R and from figure 2.19.

b) If $a = b$ and $c \ll a; c \ll b$, the radius of the circumscribed circle is equal to $R \approx a/2 = b/2$. Prove this property from the above equation for R and explain it from figure 2.19.

27*. The depressed cubic equation is given by

$$x^3 + px + q = 0. \tag{2.140}$$

Its three solutions can be expressed through trigonometric functions (see section A.29):

$$x_k = 2\sqrt{-\frac{p}{3}} \cos\left(\frac{1}{3}\cos^{-1}\left(\frac{3q}{2p}\sqrt{-\frac{3}{p}}\right) - \frac{2\pi k}{3}\right), \tag{2.141}$$

where $k = 0, 1, 2$. For the limiting case $q \to 0$, equation (2.140) factorizes to

$$x(x^2 + p) = 0, \tag{2.142}$$

and we can see that one of its roots must be equal to zero and the other ones are equal to $\pm\sqrt{-p}$. Show that these limiting cases also apply to solution (2.141).

28*. Section A.32 gives a basic explanation of the so-called Kalman filter. It is an algorithm that computes an estimate X for a quantity that is measured by two (possibly different) instruments. Suppose that the first measurement produced a value X_1 with a variance of the measurement error σ_1^2, and the second measurement produced a value X_2 with a variance of the measurement error σ_2^2. Then the best estimate for the X from these two measurements is given by the following equation:

$$X = \frac{X_1\sigma_2^2 + X_2\sigma_1^2}{\sigma_2^2 + \sigma_1^2}. \tag{2.143}$$

The accuracy of X is characterized by its own variance:

$$\sigma^2 = \frac{\sigma_2^2\sigma_1^2}{\sigma_2^2 + \sigma_1^2}. \tag{2.144}$$

Read section A.32, explore the following limiting cases, and explain the results:

a) Instrument 1 is much more accurate than instrument 2: $\sigma_1^2 \ll \sigma_2^2$.

b) Both instruments produced the same value: $X_1 = X_2$.

c) Show that for nonzero variances σ_1^2, σ_2^2, the value of X is always between the values of X_1 and X_2. Why do you expect this to be true?

d) Show that $\sigma^2 < \sigma_1^2$, $\sigma^2 < \sigma_2^2$. Why do you expect this to be true?

29*. Section A.31 provides a formula for the payment rate on a mortgage. Given the initial loan amount D, interest rate r, and duration of the loan T, the annual payment rate is

$$p \approx \frac{rDe^{rT}}{e^{rT} - 1}. \tag{2.145}$$

Explore the following limiting cases and explain the results:

a) A very short loan duration or a low interest rate: $rT \ll 1$

b) A very long loan duration or a high interest rate: $rT \gg 1$

(Hint: Use the following properties of the exponent: $e^x \approx 1 + x$ for $|x| \ll 1$ and $e^x \gg 1$ for $x \gg 1$.)

3 Symmetry

You've heard this story before: Isaac Newton saw an apple falling from a tree, and it gave him pause. Why does an apple always fall exactly downward, which is in the direction to the Earth center? And if Earth draws the apple, does the apple draw Earth? Newton's answers to these questions started a revolution in science.

Let us look closer at Newton's reasoning. The first question, why the apple falls toward the Earth center, implicitly assumes that unless there is some external factor, all directions in our physical space must be equal. Indeed, consider Earth and the apple as two nearly spherical bodies. A pair of spheres makes just one direction special: the direction along the line that connects the centers of these two spheres. An apple falling in *any other* direction would mean arbitrarily "choosing" that direction over myriad other, equally legitimate directions. Figure 3.1 illustrates this argument: direction A is unique and distinct from any other direction. Any other direction is not unique; for example, symmetric directions B and C would be equally legitimate for a falling apple. An apple falling in direction B would raise the question, why isn't it falling in direction C? and vice versa. (Note that direction A does not have a symmetric counterpart that would raise such a question.)

The second question asks if Earth and the apple must be interchangeable in a mathematical model for the interaction of these two bodies.

Both of Newton's questions are based on the notion that is the subject of this chapter: symmetry. Both were also a starting point for his earthshaking discoveries.

Symmetry is omnipresent in the world around us. Our bodies are approximately symmetric: the left half is a mirror image of the right half. A round soda can is symmetric: rotating it around its axis does not change its shape. A snowflake is symmetric: rotating it by 60° does not change the way it looks. For a square table, a rotation must be by 90° to have no effect.[1]

---

1. Symmetry is not limited to the physical world. In visual art and in architecture, symmetry is ubiquitous. It shows up in a wreath of sonnets. Palindromes read forward and backward in the same way, except for the white space and punctuation (try "A man, a plan, a canal—Panama!" by Leigh Mercer). Even more amazing are mirror canons in music.

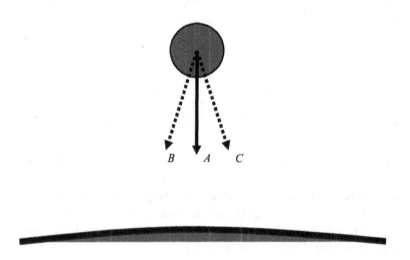

Figure 3.1
Symmetry in the fall of Newton's apple

Each of these examples has two key attributes:

1. There is an operation on the object, such as a reflection or rotation.
2. This operation either leaves the object intact or changes it in a predictable way. In the first case, we say that the object is *invariant* with respect to that operation. If the object is not invariant with respect to an operation, but the change is predictable, we still can use symmetry for analysis.

For Newton's apple, symmetry means there is only one special direction in space: to Earth's center. There are no other preferred directions because the Earth-apple system is symmetric (invariant) with respect to a rotation around the line connecting these two bodies. A rotation with respect to this line leaves the Earth-apple system intact. The same is true for a left-right reflection (see figure 3.1). An apple falling in any direction other than toward the Earth center would violate that symmetry. This second symmetry (and related invariance) deals with the swap of Earth and the apple. Both are physical bodies, and the laws of physics should apply equally to them. If Earth attracts objects (such as apples), then the apple must attract Earth. But then *all* objects must attract each other. The law of universal gravitation is born! (Note the "universal" qualifier here – it applies to planets, apples, and elementary particles alike.)

The concept of symmetry proves to be extremely fruitful for problem solving. In this chapter we look at how to find and use symmetry in mathematical problems.

3.1 Symmetry in Mathematical Problems

In mathematics, the concept of symmetry was formalized by Évariste Galois. He started an entirely new area of mathematics: group theory. While group theory finds important applications in mathematics and in physics, it is beyond the scope of this book. We will limit our exploration of this subject to a few simple, intuitive tools that may help in solving various practical problems.

A problem has an intrinsic symmetry when a certain controlled change in the formulation of the problem does not produce a change in the solution. In this case, as noted above, we say that the solution is *invariant* with respect to that particular change in the formulation of the problem. In some other cases, the solution does change, but the change is predictable from some extraneous considerations. Both situations offer you an opportunity to test your result and to understand it better, much like a limiting case does.

For example, let's look again at the two hikers problem. Consider two hikers starting to walk toward each other from the opposite ends of a trail (see figure 3.2). One hiker maintains the speed V_1, and another maintains the speed V_2. The total length of the trail is D. What distances will they cover before meeting on the trail?

Section A.1 shows that hikers 1 and 2 will travel the following distances:

$$
\begin{aligned}
D_1 &= \frac{DV_1}{V_1 + V_2}, \\
D_2 &= \frac{DV_2}{V_1 + V_2}.
\end{aligned}
\tag{3.1}
$$

Think about what happens if the speeds of the hikers are swapped compared to the original formulation. Assume that the first hiker now walks with the speed V_2, and the second

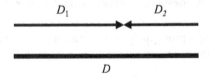

Figure 3.2
Two hikers on a trail: symmetry

hiker walks with the speed V_1. Then we expect them to meet at a point that is the mirror image of the meeting point in the original solution. This is equivalent to swapping D_1 and D_2.

The symmetry here means that if we swap V_1 and V_2 *and* D_1 and D_2 in the equations, the equations must remain valid. Indeed, if we swap $V_1 \leftrightarrow V_2$ and $D_1 \leftrightarrow D_2$ in equations (3.1), the first equation becomes swapped with the second one, but the *system* of these two equations remains intact.

Had we made a mistake in our solution, we could have arrived at a wrong result, for example, the following formulas:

$$\text{✎} \quad D_1 = \frac{DV_1}{V_1 - V_2},$$
$$D_2 = \frac{DV_2}{V_1 - V_2} \quad \text{✍}. \tag{3.2}$$

In this case, the symmetry check would quickly point us to an error. A swap $V_1 \leftrightarrow V_2$ and $D_1 \leftrightarrow D_2$ would flip the signs of the distances traveled, which may not be right. This is an example of how a symmetry consideration can check your solution for errors.

Just as is the case with units or limiting cases, a passed symmetry check does not guarantee that a formula is correct. (Note that our "wrong solution" (3.2) does have the units right!) But if your result does *not* pass a symmetry check, something must be wrong.

Use of symmetry is not limited to analyzing already solved problems. It can help you look ahead for the right form of the solution and to chart a better course for getting the answer.

A common trick is to choose a new variable or variables that make equations symmetric or that conform to the symmetry already present in the system. For example, if you are building a mathematical model for the waves that are produced by dropping a stone in a pond, you may want to use cylindrical (rather than Cartesian) coordinates—this exploits the axial symmetry of the waves (figure 3.3A). However, if you are building a model for the ocean tide that rolls on a straight ocean beach, Cartesian coordinates would work better (figure 3.3B).

To illustrate how symmetry can help with solving a problem, let us consider the following trigonometric equation that we need to solve for x:

$$\sin(x - a) + \sin(x - b) = c. \tag{3.3}$$

Use of a trigonometric identity for the sine of a difference[2] produces an equation that contains both sines and cosines of x:

---

2. The trigonometric identity for the sine of a sum or a difference: $\sin(p \pm q) = \sin p \cos q \pm \cos p \sin q$.

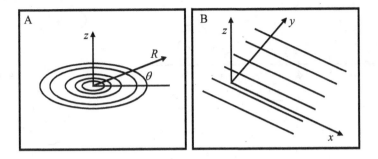

Figure 3.3
Waves in a pond and on a beach

$$\sin x \cos a - \cos x \sin a + \sin x \cos b - \cos x \sin b = c. \tag{3.4}$$

This equation can be reduced to one that is solved in section A.13. That particular solution requires a "lucky guess" that may not at all be obvious. If you do not have the insight to come up with such a guess, there is another way to deal with the original equation (3.3).

Note that there are two parameters in the left-hand side: one shifts the sine function by a, and another shifts another sine function by a (generally) different value b. Let us see what happens if these shifts are *symmetric*. Here we seek to impose a symmetry on the original equation, hoping it will lead to a breakthrough.

Key Point

It often helps to make equations symmetric by choosing variables smartly.

We define a new variable y that is produced by shifting the original unknown x so that the point $y = 0$ is located exactly at the midpoint between parameters a and b:

$$y = x - \frac{a+b}{2}. \tag{3.5}$$

The relationship between x and y is illustrated in figure 3.4. Points a and b lie on both sides and at the same distance from point $y = 0$. Therefore, the change of variable from x to y introduces a symmetry in the problem. It is this symmetry that will simplify the solution.

We solve equation (3.5) for x and substitute the result in equation (3.3). This yields

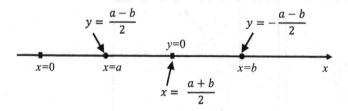

Figure 3.4
Changing a variable to impose symmetry

$$\sin(x - a) + \sin(x - b) =$$
$$\sin\left(y + \frac{a+b}{2} - a\right) + \sin\left(y + \frac{a+b}{2} - b\right) =$$
$$\sin\left(y - \frac{a-b}{2}\right) + \sin\left(y + \frac{a-b}{2}\right) = c. \tag{3.6}$$

We have a double occurrence of an expression here, and it is convenient to introduce a new variable for it; this will make the equations more compact. We denote

$$d = \frac{a-b}{2}. \tag{3.7}$$

Using this new notation, we then apply the same trigonometric identity for the sine of a sum or a difference to get

$$\sin(y - d) + \sin(y + d) =$$
$$\sin y \cos d - \cos y \sin d + \sin y \cos d + \cos y \sin d = c. \tag{3.8}$$

The terms with $\cos(y)$ cancel out, and we get

$$2 \sin y \cos d = c. \tag{3.9}$$

This equation is much easier to solve: assuming that $\cos d \neq 0$, we divide both parts by $2 \cos d$ and compute the inverse sine of both parts. This gives an expression for y, and then we go back to the original variable x:

$$y = (-1)^n \sin^{-1}\left(\frac{c}{2\cos\frac{a-b}{2}}\right) + \pi n,$$

$$x = y + \frac{a+b}{2},$$

(3.10)

where n is an integer.

The takeaway message from this example is that a clever change of variables leads to a simplification of the problem: we are left with an expression for only $\sin(x)$, and the terms with $\cos(x)$ are eliminated. In this case, the change of variables makes the equation symmetric, and the symmetry makes it simpler to solve.

To summarize, having symmetry in your problem is a good thing: it allows you to apply checks to your solution and may make calculations simpler.

Now we proceed to the application of symmetry analyses in more detail.

3.2 The Product of Two Linear Expressions

In chapter 1 we saw that units must be consistent not just at the final result but throughout the entire solution. The same is true for symmetry and invariance. For example, let's look at the problem in section A.7:

$$(x - a)(x - b) = d. \tag{3.11}$$

The solution of this problem is given by

$$x_{1,2} = \frac{(a+b) \pm \sqrt{(a-b)^2 + 4d}}{2}. \tag{3.12}$$

Note that parameters a and b enter equation (3.11) symmetrically. Suppose we first solve it for a pair of values of a and b, and then for a new pair of values $a_{\text{new}} = b$ and $b_{\text{new}} = a$. An inspection of equation (3.11) shows that swapping the values of a and b does not alter it. If this fact is true for the original problem, it must be true for the solution as well: the solution must be in such a form that a swap of variables a and b does not alter the value of x. Indeed, parameters a and b enter solution (3.12) twice: as a sum $a + b$ and as a squared difference $(a-b)^2$. Obviously, the value of $a+b$ is not affected by a swap of a and b. While the swap flips the sign of the difference $a - b$, it does not change the value of $(a-b)^2$. We see that the invariance with respect to the swap of a and b applies not just to the original equation but also to its solution.

Just as with limiting cases, symmetry and invariance create a bridge between the start and the end points of our solution. Similar to the application of limiting cases, symmetry is used to check the solution and to get additional insight. For example, imagine that we made an error in solving equation (3.11). Below we repeat a solution from section A.7 but introduce an error in one of the intermediate steps.

Key Point

Analysis of symmetry is a great tool to find errors in your solution.

We start from equation (3.11), where we expand the expression in the left-hand side and regroup the terms:

$$x^2 - x(a + b) + ab - d = 0. \tag{3.13}$$

This is a quadratic equation with the discriminant $D = (a + b)^2 - 4a(b - d)$. The solution for x is given by the quadratic formula:

$$x_{1,2} = \frac{(a - b) \pm \sqrt{(a + b)^2 - 4a(b - d)}}{2}. \tag{3.14}$$

This solution does *not* have symmetry with respect to swapping a and b because

$$(a + b)^2 - 4a(b - d) \neq (b + a)^2 - 4b(a - d) \tag{3.15}$$

Yet the symmetry with respect to $a \leftrightarrow b$ is present in the original equation. We immediately conclude that equation (3.14) is wrong. Moreover, by looking at the preceding steps, we see that equation (3.13) still possesses the required symmetry. This tells us that we must have made an error in getting from equation (3.13) to equation (3.14). A closer inspection shows that the error is in the expression for the discriminant. This is an example of how symmetry analysis helps us pinpoint an error in a derivation. (Please refer to section A.7 for the correct solution of this problem.)

3.3   The Intersections between a Circle and a Straight Line

In previous chapters, we already have analyzed the problem of finding intersections between a circle and a straight line. A circle and a line are given by the following equations:

$$x^2 + y^2 = R^2,$$
$$y = px + q. \tag{3.16}$$

Horizontal coordinates $x_{1,2}$ of the intersections (if any) are given by

$$x_{1,2} = \frac{-pq \pm \sqrt{(1 + p^2) R^2 - q^2}}{1 + p^2}, \tag{3.17}$$

where subscripts $1, 2$ correspond to the $\pm$ signs in the right-hand side. The vertical coordinates of the intersections are obtained by plugging $x_{1,2}$ in the second equation of system (3.16). As a reminder, R denotes the radius of the circle, p is the slope of the straight line, and q is the point of its intersection with the vertical axis. The full solution is given in section A.3.

Here we revisit this problem to explore any symmetries in it. Some of these symmetries are related to limiting cases, which we investigated in section 2.5. Exploring multiple symmetries provides us with a multipoint view of this problem:

1. A zero slope case. Here we consider the case of $p = 0$. In this case, the line is horizontal, and its intersections with the circle should be symmetric with respect to the vertical axis. Therefore, we should expect to get $x_1 = -x_2$ and $y_1 = y_2$. This case is shown in figure 3.5A.

    Indeed, if we plug $p = 0$ in solution (3.17), we get

    $$x_{1,2} = \pm \sqrt{R^2 - q^2}. \tag{3.18}$$

    We do get $x_1 = -x_2$. The second equation of system (3.16) is reduced to

    $$y_{1,2} = q. \tag{3.19}$$

    As expected, $y_1 = y_2$. This symmetry condition holds.

2. A zero vertical offset case. Now we consider the case of $q = 0$, that is, when the straight line crosses the vertical axis at the origin. This case is illustrated in figure 3.5B. We see that we should get $x_1 = -x_2$ and $y_1 = -y_2$.

    Plugging $q = 0$ in equation (3.17) yields

    $$x_{1,2} = \pm \frac{R}{\sqrt{1 + p^2}}. \tag{3.20}$$

    If we use $q = 0$ and these values of $x_{1,2}$ in the second equation of system (3.16), we get

    $$y_{1,2} = \pm \frac{pR}{\sqrt{1 + p^2}}. \tag{3.21}$$

    Indeed, we get $x_1 = -x_2$ and $y_1 = -y_2$. This symmetry condition also holds.

3. Now we combine the previous two conditions and require *both* p and q to be zero. On a drawing, the straight line would match the x axis. In this case, both above symmetries must apply. We still have $x_1 = -x_2$ here because this condition follows from both symmetries. For the vertical coordinates, the situation is different. The first symmetry mandates that $y_1 = y_2$, while the second symmetry mandates that $y_1 = -y_2$. The only way to satisfy both conditions is to have $y_1 = y_2 = 0$, which is indeed the case here. This shows how multiple symmetry constraints can work together.

4. Next, we explore a change in the signs of the parameters. We first look at the symmetry that arises from $p \leftrightarrow -p$, that is, from a change in the sign of the slope of the line. This

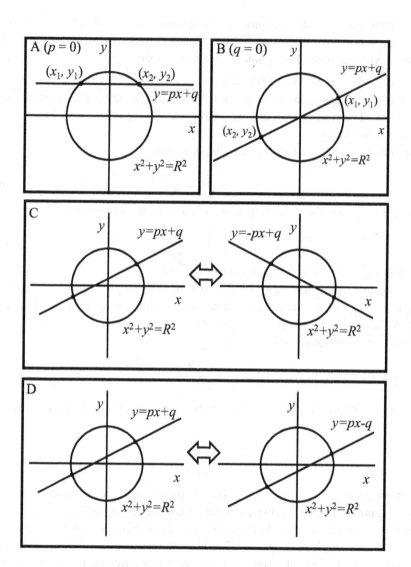

Figure 3.5
A circle and a straight line: symmetry

case is shown in figure 3.5C. In the figure, we see that this operation should change the signs of x_1, x_2 but retain the values of y_1, y_2. Indeed, plugging $p' = -p$ in equation (3.17) does produce the following transformation for $x_{1,2}$:

$$
\begin{aligned}
x_1' &= -x_2, \\
x_2' &= -x_1.
\end{aligned}
\tag{3.22}
$$

This result does exhibit the expected symmetry, with the caveat that the symmetry requires also swapping the two roots. The last requirement is immaterial. We numbered the roots by using the plus sign for x_1 and the minus sign for x_2, but this choice was arbitrary, and the result must not depend on it. By using $p' = -p$ and equations (3.22) in the second equation of (3.16), we confirm that values of $y_{1,2}$ remain invariant.

Key Point

Whenever you make an arbitrary choice in defining or selecting variables between two or more equal options, look for symmetry in the equations with respect to these options or deem the effect of swapping such options to be immaterial.

---

5. A change in the sign of q is illustrated in figure 3.5D. We expect to see a flip in the signs of both $x_{1,2}$ and $y_{1,2}$. Indeed, using $q' = -q$ in equation (3.17) again produces the same result as in equations (3.22). However, a substitution in the second equation of (3.16) now flips the signs of $y_{1,2}$ as well.

6. The last symmetry is a bit trickier. Let us solve the second equation of system (3.16) for x, assuming that $p \neq 0$:

$$
x = \frac{1}{p}y - \frac{1}{p}q.
\tag{3.23}
$$

This equation has the same form as the second equation in (3.16) if we make the following replacements, essentially swapping x and y:

$$
\begin{aligned}
x'' &= y, \\
y'' &= x, \\
p'' &= \frac{1}{p}, \\
q'' &= -\frac{1}{p}q.
\end{aligned}
\tag{3.24}
$$

In addition, the first equation in (3.16) is symmetric with respect to swapping x and y and does not contain p or q. It remains unaffected by the transformation of variables that is specified by equations (3.24). With these observations in mind, we expect the solution to be invariant with respect to the replacements given by equations (3.24). Let us check if this is the case. We use x'', y'', p'', and q'' in equation (3.17) and substitute their values from equations (3.24):

$$x''_{1,2} = \frac{-p''q'' \pm \sqrt{(1 + (p'')^2) R^2 - (q'')^2}}{1 + (p'')^2}$$

$$= \frac{\frac{q}{p^2} \pm \sqrt{\left(1 + \frac{1}{p^2}\right) R^2 - \frac{q^2}{p^2}}}{1 + \frac{1}{p^2}} \tag{3.25}$$

$$= \frac{q \pm p \sqrt{(1 + p^2) R^2 - q^2}}{1 + p^2}.$$

According to our symmetry analysis, the last expression must be equal to y_1 and y_2 (depending on the choice of the sign). Indeed, computation of $y_{1,2}$ from equation (3.17) and the second equation of (3.16) yields

$$y_{1,2} = px_{1,2} + q$$

$$= p\frac{-pq \pm \sqrt{(1 + p^2)R^2 - q^2}}{1 + p^2} + q$$

$$= \frac{-p^2q \pm p \sqrt{(1 + p^2)R^2 - q^2} + q + p^2q}{1 + p^2} \tag{3.26}$$

$$= \frac{q \pm p \sqrt{(1 + p^2)R^2 - q^2}}{1 + p^2}.$$

Comparing the last lines in equations (3.25) and (3.26) shows that $x''_{1,2} = y_{1,2}$, as expected. Proving that $y''_{1,2} = x_{1,2}$ is left as an exercise for the reader.

Equations (3.24) describe a reflection symmetry that swaps the x and y axes with respect to a 45° line. Note that the transformation for p and q is more complex than a simple swap; this is attributed to the original formulation of the equation for the straight line in equation (3.16), which is asymmetric.

Key Point

Multiple symmetries can show different facets of a problem.

3.4 A Circle Inscribed in a Right Triangle

Consider a circle that is inscribed in a right triangle (see figure 3.6 and section A.27). Given a leg of a right triangle a and the adjacent angle α, the radius R of the circle that is inscribed in the triangle is given by

$$R = \frac{a \sin \alpha}{1 + \cos \alpha + \sin \alpha}. \qquad (3.27)$$

Our search for symmetry in this problem starts with an observation: the problem's formulation contains an arbitrary choice. We have selected one particular leg a and the adjacent angle α, but we could have equally used the *other* leg and *its* adjacent angle. Yet, the result for the radius of the inscribed circle would have remained the same. This symmetry can (and should) be checked in the final result. Let us see if this symmetry holds.

We denote the other leg as b, and its adjacent angle as β. If we use b in place of a and β in place of α, the functional form of solution (3.27) must remain the same. This follows from the fact that both legs of the triangle play identical roles in any derivation, and our choice of one or another must not matter. Therefore, we should expect to get the following solution for the circle radius as expressed through b and β:

$$R = \frac{b \sin \beta}{1 + \cos \beta + \sin \beta}. \qquad (3.28)$$

Values given by equations (3.27) and (3.28) must be equal because they apply to the same circle. Let us see if this is the case.

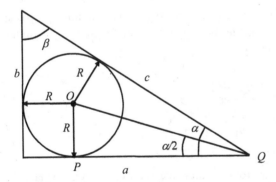

Figure 3.6
A circle inscribed in a right triangle: symmetry

Since we've got a right triangle, we know that $\beta = \pi/2 - \alpha$. We substitute this value in the denominator of equation (3.28) and make use of the appropriate trigonometric identities[3] (note that we leave the numerator intact):

$$
\begin{aligned}
R &= \frac{b \sin \beta}{1 + \cos\left(\frac{\pi}{2} - \alpha\right) + \sin\left(\frac{\pi}{2} - \alpha\right)} \\
&= \frac{b \sin \beta}{1 + \sin \alpha + \cos \alpha}.
\end{aligned}
\tag{3.29}
$$

Comparing this equation with equation (3.27), we see that they produce the same value for R if

$$
b \sin \beta = a \sin \alpha.
\tag{3.30}
$$

This last condition is just the law of sines!

The important takeaway from this problem is that the result must not depend on any arbitrary choice that we have made in formulating or solving a problem. In this case it was a choice of one leg of the triangle versus another; in other problems, it may be a choice of variables or coordinate system axes. In any case, the solution must be *invariant* with respect to that choice. Therefore, the solution must possess the symmetry that corresponds to that invariance.

3.5   Blending Syrups

Consider three syrups with masses m_1, m_2, and m_3 and sugar concentrations p_1, p_2, and p_3. Sections A.16 and A.17 show that the blends of two or three of these syrups will have sugar concentrations respectively given by

$$
\begin{aligned}
p_{12} &= \frac{p_1 m_1 + p_2 m_2}{m_1 + m_2}, \\
p_{123} &= \frac{p_1 m_1 + p_2 m_2 + p_3 m_3}{m_1 + m_2 + m_3}.
\end{aligned}
\tag{3.31}
$$

These problems have three symmetries:

1. We check if swapping syrups 1 and 2 (or any pair in the case of three syrups) makes any difference. A look at equations (3.31) shows that they remain invariant. This symmetry holds.

2. Next, we check what happens if we blend two syrups first and then add the third one, as compared to mixing all three syrups at the same time. We expect the outcome not

---

3. Trigonometric identities: $\sin(\pi/2 - x) = \cos x$; $\cos(\pi/2 - x) = \sin x$.

to depend on how we mix the ingredients. The mixing order is an arbitrary choice that we make, and section 3.3 noted that a solution must be invariant with respect to such a choice. We can check this invariance in the solution.

Blending two syrups first produces a mix with the concentration that is given by the first equation in (3.31). The resulting mix is blended with the third syrup. This is another blending of two syrups. It must be described by the same equation, but with new inputs. Therefore, we use the formula for the two-syrup problem twice: first to produce the value p_{12}, and then we use this value again as an input, along with p_3, to compute the concentration of the three-syrup mix:

$$
\begin{aligned}
p'_{123} &= \frac{p_{12}m_{12} + p_3 m_3}{m_{12} + m_3} \\
&= \frac{\frac{p_1 m_1 + p_2 m_2}{m_1 + m_2}(m_1 + m_2) + p_3 m_3}{(m_1 + m_2) + m_3} \\
&= \frac{p_1 m_1 + p_2 m_2 + p_3 m_3}{m_1 + m_2 + m_3}.
\end{aligned}
\tag{3.32}
$$

We do get the same value as in the second equation of (3.31). This confirms that blending two syrups first and adding the third one later is equivalent to mixing all three together, as it should be.

3. Below we illustrate the final symmetry for the two-syrup version of the problem. The same symmetry for a three-syrup mix is left as an exercise.

Concentrations p_1 and p_2 are defined by selecting one particular ingredient (sugar) from the mix. This selection is an arbitrary choice, and our results should not depend on it. Therefore, we should get the same equations if we consider the concentration of the water (instead of the sugar) in the syrup as our variable. We denote the values for the concentrations of water as q_1 and q_2. Since the concentrations of all the ingredients must sum to one, we get

$$
\begin{aligned}
q_1 &= 1 - p_1, \\
q_2 &= 1 - p_2, \\
q_{12} &= 1 - p_{12}.
\end{aligned}
\tag{3.33}
$$

The first equation of (3.31) must be valid for concentrations q_1, q_2, and q_{12}, just as it is valid for p_1, p_2, and p_{12}:

$$
q_{12} = \frac{q_1 m_1 + q_2 m_2}{m_1 + m_2}.
\tag{3.34}
$$

Substituting q_1, q_2, and q_{12} here yields

$$1 - p_{12} = \frac{(1 - p_1)m_1 + (1 - p_2)m_2}{m_1 + m_2}. \tag{3.35}$$

This transforms to

$$1 - p_{12} = \frac{m_1 + m_2 - p_1 m_1 - p_2 m_2}{m_1 + m_2}, \tag{3.36}$$

which does reduce to the first equation in (3.31), as required.

We see that a seemingly simple problem exhibits multiple symmetries, and this is in addition to several illuminating limiting cases that we have already investigated above. This wealth of features helps us understand the problem better.

3.6 Draining a Pool

Consider a pool that must be drained for winter. There are three pumps that can be used separately or jointly. The individual pumps can drain this pool in T_1, T_2, and T_3 hours. Sections A.18 and A.19 show that if we use two or three pumps jointly, the amount of time to drain the pool is respectively given by

$$T_{12} = \frac{T_1 T_2}{T_1 + T_2},$$
$$T_{123} = \frac{T_1 T_2 T_3}{T_1 T_2 + T_1 T_3 + T_2 T_3}. \tag{3.37}$$

These equations feature two symmetries.

In equations (3.37), we can swap the values of any pair of T_1, T_2, and T_3, and the result will remain the same. For example, if we swap $T_1 \leftrightarrow T_2$, neither equation changes. This symmetry is to be expected: our choice for the numbering of pumps is arbitrary, and the result must not depend on it.

In addition to this symmetry, there is a less obvious one. Let us consider a pool that is being drained by three pumps, but now we mentally group pumps 1 and 2 into a more powerful "virtual" pump. In other words, we consider the problem of draining a pool with two pumps, where one of them is in turn an aggregate of two smaller pumps.

Grouping pumps 1 and 2 alters our way of thinking about the problem but does not affect the physical process of draining the pool. Similar to any arbitrary choice that we make when solving a problem (see section 3.3), this thought experiment should leave the final result intact. For the problem in hand, we expect that the result should remain invariant with respect to any grouping of the pumps. Let us see if this is the case.

To check this invariance, we first need to determine the productivity of the virtual pump formed by two individual pumps. The time T_{12} that is required to drain the pool using pumps 1 and 2 is given by the first equation in (3.37).

In the next step, these two pumps are considered as one composite virtual pump that in turn works with pump 3. Draining the pool is now again described by the two-pump equation, where we use durations T_{12} and T_3 as inputs:

$$T'_{123} = \frac{T_{12}T_3}{T_{12} + T_3},$$
(3.38)

where T'_{123} denotes the time that is required to drain the pool using all three pumps when two of them are aggregated in a larger "virtual" pump. Substitution of T_{12} from equations (3.37) yields

$$
\begin{aligned}
T'_{123} &= \frac{\frac{T_1 T_2}{T_1 + T_2} T_3}{\frac{T_1 T_2}{T_1 + T_2} + T_3} \\
&= \frac{\frac{T_1 T_2 T_2}{T_1 + T_2}}{\frac{T_1 T_2 + (T_1 + T_2)T_3}{T_1 + T_2}} \\
&= \frac{T_1 T_2 T_3}{T_1 T_2 + T_1 T_3 + T_2 T_3}.
\end{aligned}
$$
(3.39)

We did get the same equation as in the original solution (3.37) for three pumps! This shows that our thought experiment about combining two pumps into one is valid, as expected.

Key Point

Sometimes a problem allows multiple approaches or formulations that produce the same result. Look for the invariance that is linked to the equivalence of these approaches or formulations.

For comparison, let us see what would happen if we made a mistake in the problem with two pumps. Suppose that we have reached the following erroneous result:

$$\tilde{T}_{12} = \frac{2T_1 T_2}{T_1 + T_2}.$$
(3.40)

For this result, simple checks of units and the symmetry with respect to swapping T_1 and T_2 do not catch the error. To see that this formula is invalid, we would need to perform a limiting case analysis (section 2.10) or to look at the invariance with respect to the "aggregation" of two pumps. Following the same logic, we can get a result for three pumps when using the erroneous formula above:

$$\tilde{T}'_{123} = \frac{2\frac{2T_1T_2}{T_1+T_2}T_3}{\frac{2T_1T_2}{T_1+T_2} + T_3}$$

$$= \frac{4T_1T_2T_3}{2T_1T_2 + T_1T_3 + T_2T_3}. \qquad (3.41)$$

We can see that this result is wrong: it is not symmetric with respect to swapping T_1 and T_3 or T_2 and T_3. This is another case when a symmetry argument helps catch an error.

3.7   The Sum of an Unknown and Its Reciprocal

Here we turn to the problem for a sum of an unknown and its reciprocal:

$$x + \frac{1}{x} = d. \qquad (3.42)$$

The solution is provided in section A.8:

$$x_{1,2} = \frac{d \pm \sqrt{d^2 - 4}}{2}, \qquad (3.43)$$

where subscripts $1, 2$ correspond to the $\pm$ signs in the right-hand side. For this problem, we explore two symmetries:

1. We see that this problem has two solutions, denoted here as x_1 and x_2. It also has a symmetry with respect to swapping x and $1/x$ in the original equation (3.42). Unless $x_1 = \pm 1$, replacement of that root with its reciprocal ($x_1 \leftrightarrow 1/x_1$) will produce a different value, and the symmetry dictates that this new value also must be a root of equation (3.42). Since there are only two roots, the only possibility for $1/x_1$ to be different from x_1 and still be a root of equation (3.42) is to have $1/x_1 = x_2$. Therefore, the symmetry requires the two roots of this equation to be reciprocals of each other:

$$x_1 \cdot x_2 = 1. \qquad (3.44)$$

    Let us check if this is indeed the case. We compute the product of the two roots given by equation (3.43) and use the formula for the product of a sum and a difference of two variables[4] to simplify the result:

---

4. The formula for the product of a sum and a difference of two variables: $(a + b)(a - b) = a^2 - b^2$.

$$x_1 \cdot x_2 = \frac{d + \sqrt{d^2 - 4}}{2} \cdot \frac{d - \sqrt{d^2 - 4}}{2}$$

$$= \frac{d^2 - \left(\sqrt{d^2 - 4}\right)^2}{4}$$

$$= \frac{d^2 - (d^2 - 4)}{4} \tag{3.45}$$

$$= 1.$$

The roots are reciprocal, as expected.

2. Another symmetry in this problem is a recurring theme throughout this chapter: we explore changing the sign of a parameter. If we change the sign of d in equation (3.42), we see that it remains valid if we also change the sign of x in the left-hand side. The same must be true for solution (3.43). If we change $d \leftrightarrow -d$, we will get

$$x'_{1,2} = \frac{-d \pm \sqrt{d^2 - 4}}{2}. \tag{3.46}$$

We see that

$$x'_1 = -x_2,$$
$$x'_2 = -x_1. \tag{3.47}$$

The roots are swapped here as well. This is immaterial because our choice for root numbering is arbitrary and the result must have a symmetry for root swapping.

3.8   Designing Satellite Coverage

In this section we revisit the problem for designing satellite coverage. A satellite flies at altitude H above Earth and transmits a signal downward. Figure 3.7 shows a cross section of this problem. The satellite beam covers a round spot on Earth's surface whose diameter spans the distance between points B and C. The antenna beam has a conical shape, and the angle from the center to the edge of the beam is α. Earth's radius is R. Section A.24 computes the length of the arc on Earth's surface from the center of the beam spot to its edge:

$$L = R\left(\sin^{-1}\left(\frac{R + H}{R} \sin \alpha\right) - \alpha\right). \tag{3.48}$$

Let's see what happens if we flip the sign of α. Substitution $\alpha' = -\alpha$ in equation (3.48) yields

$$L(-\alpha) = R\left(-\sin^{-1}\left(\frac{R + H}{R} \sin \alpha\right) + \alpha\right) = -L(\alpha). \tag{3.49}$$

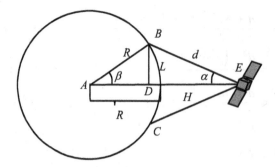

Figure 3.7
Designing satellite coverage: symmetry

We see that L is an odd function[5] of α. This has an intuitive interpretation: flipping the sign of α flips the top and bottom parts of figure 3.7. If L is measured using angle β, and the sign of β is flipped along with the sign of α, we expect L to have a different sign too. In some other solution of this problem, L could have been determined as a Euclidean distance using the Pythagorean theorem; in this case, we would get an even function of angle α. But we do not expect to get a result that is neither an odd nor an even function of α.

Imagine that we made a mistake in obtaining equation (3.48) and got the following wrong result:

$$\tilde{L}(\alpha) = R\left(\sin^{-1}\left(\frac{R+H}{R}\sin\alpha\right) - \alpha^2\right). \tag{3.50}$$

This solution would lack the symmetry for changing the sign of α because it produces a function that is neither odd nor even:

$$\tilde{L}(-\alpha) = R\left(-\sin^{-1}\left(\frac{R+H}{R}\sin\alpha\right) - \alpha^2\right) \neq \pm\tilde{L}(\alpha) \quad . \tag{3.51}$$

This would serve as an indication that this solution is in error.

The very definition of even and odd functions is based on symmetry. This problem shows that a mathematical symmetry for an odd function can be linked to a real-life symmetry. Exercise 18 at the end of this chapter contains an example of a real-life symmetry that leads to an even function in the solution.

---

5. By definition, an odd function $F(x)$ has a domain that is symmetric with respect to $x = 0$ and the function obeys the property $F(-x) = -F(x)$. An even function also has a domain that is symmetric with respect to $x = 0$, but the function obeys the property $F(-x) = F(x)$. For example, $\cos x$ is an even function ($\cos(-x) = \cos x$), while $\sin x$ is an odd function ($\sin(-x) = -\sin x$).

Key Point

In some cases, the problem setup may dictate that your solution must be an even or an odd function of a parameter. Changing the signs of various parameters is a great way to explore potential symmetries.

3.9 Two Circles Inscribed in an Angle

Consider the problem for two circles that are inscribed in an angle in such a way that they touch each other (figure 3.8A). Section A.26 computes the ratio of the radii of these two circles:

$$\frac{R}{r} = \frac{1 + \sin \frac{\beta}{2}}{1 - \sin \frac{\beta}{2}}. \tag{3.52}$$

What happens if the angle goes negative? Does this case make any sense? To investigate this, consider small values of β (limiting case 1 in section 2.13). From that, we will take baby steps to the zero-angle case and then to a small negative-angle case.

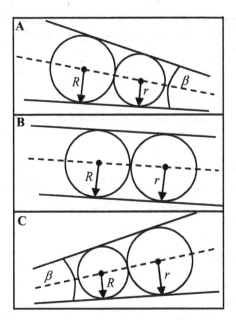

Figure 3.8
Two circles inscribed in an angle: symmetry

Figure 3.8 shows all three cases in separate panes A, B, and C. This visualization helps explain the change in angle β as a pivot of the top line around a fixed point in the middle. As we vary angle β, the lines become parallel for $\beta = 0$. For $\beta > 0$ and $\beta < 0$, the lines converge on the opposite sides of the figure. For consistency, we denote the radius of the circle on the left as R and the radius of the circle on the right as r, even though this notation no longer associates the larger circle with the uppercase letter.

The last statement suggests an interesting observation. For negative angles (figure 3.8C), we see that $R < r$; moreover, we should see a symmetry here: a change in the sign of β should swap the values of R and r. Indeed, if we look at equation (3.52) and substitute $\tilde{\beta} = -\beta$, the denominator and numerator do get swapped:

$$
\frac{\tilde{R}}{\tilde{r}} = \frac{1 + \sin \frac{\tilde{\beta}}{2}}{1 - \sin \frac{\tilde{\beta}}{2}} =
$$
$$
\frac{1 + \sin \frac{-\beta}{2}}{1 - \sin \frac{-\beta}{2}} = \tag{3.53}
$$
$$
\frac{1 - \sin \frac{\beta}{2}}{1 + \sin \frac{\beta}{2}} = \frac{r}{R}.
$$

In the original formulation of this problem, we may have assumed that angle measures are positive. Symmetry took us beyond this limitation and into a new domain for the problem, but the results still make sense there!

Key Point

Exploration of symmetry can point to a new domain of applicability for the problem.

3.10   The Sum or Difference of Two Radicals

In this section we show how symmetry helps identify a subtle error in a solution. We consider two problems that differ just by a sign in front of one term. Both can be generalized to the following equation for x where $a > 0$:

$$
\sqrt{a + x} \pm \sqrt{a - x} = b. \tag{3.54}
$$

A solution to this generalized problem is provided below. The two versions corresponding to the plus and the minus sign in equation (3.54) are solved separately in sections A.20 and A.21. We do not refer to these separate derivations here; instead, we purposely present a solution that combines both versions in one but contains a subtle oversight. This oversight leads to an issue with the final result. We will use symmetry analysis to identify this issue.

To solve for x, we square equation (3.54) to get

$$2a \pm 2\sqrt{a^2 - x^2} = b^2. \tag{3.55}$$

Subtracting $2a$ from both sides and another squaring of this equation yields

$$a^2 - x^2 = \left(\frac{b^2}{2} - a\right)^2. \tag{3.56}$$

The final solution for x is as follows:

$$x_{1,2} = \pm\sqrt{a^2 - \left(\frac{b^2}{2} - a\right)^2}, \tag{3.57}$$

where subscripts $1, 2$ correspond to the $\pm$ signs in the right-hand side of equation (3.57) but are not linked with the $\pm$ signs in the original equation (3.54).

Let us now investigate if the original equation and the final solution have any symmetries. First, we look at the version of the original equation with the plus sign:

$$\sqrt{a + x} + \sqrt{a - x} = b. \tag{3.58}$$

We see that if x_1 is a root of this equation, then $x_2 = -x_1$ must also be a root. The equation is therefore symmetric with respect to changing the sign of x. Indeed, our solution (3.57) satisfies this condition, as it produces both signs for the roots.

Next, we turn to the version with the minus sign:

$$\sqrt{a + x} - \sqrt{a - x} = b. \tag{3.59}$$

In this case, flipping the sign of x does *not* satisfy the same equation because it flips the sign of the left-hand side. Indeed, if $\tilde{x} = -x$, then

$$\sqrt{a - \tilde{x}} - \sqrt{a + \tilde{x}} = b. \tag{3.60}$$

Equivalently,

$$\sqrt{a + \tilde{x}} - \sqrt{a - \tilde{x}} = -b. \tag{3.61}$$

There is a symmetry here, but it requires flipping both the sign of x *and* the sign of b. However, the final result (3.57) is still consistent with the symmetry for flipping the sign of x only. This points us to a possible issue with this result.

The issue (treated correctly in section A.21) arises from the careless squaring equation (3.54). For the case of a difference of two radicals in the left-hand side, b can have any sign. Squaring that equation wipes out any information on the sign of b. Section A.21 shows that a careful treatment of this problem produces the following solution:

$$x = \sqrt{a^2 - \left(\frac{b^2}{2} - a\right)^2}, \quad \text{if} \quad b \geq 0, \text{ and}$$

$$x = -\sqrt{a^2 - \left(\frac{b^2}{2} - a\right)^2}, \quad \text{if} \quad b < 0.$$

(3.62)

This solution does have the required symmetry: flipping both the sign of b and the sign of x leaves the problem, and its solution, intact. Also note that this issue does not arise for the version of the problem with the plus sign: the value of b must be nonnegative there, and squaring b does not wipe out information about its sign. This is the reason for our solution (3.57) to remain correct after squaring the original equation for the plus sign version.

Key Point

Analysis can catch even subtle errors if they violate the symmetry of the problem.

After some algebra, the solutions for both equations can be presented in a simpler way:

1. For the plus sign in equation (3.54) we get

$$x_{1,2} = \pm b \sqrt{a - \frac{b^2}{4}},$$

(3.63)

   where we must have $2\sqrt{a} \geq b \geq \sqrt{2a}$.

2. For the minus sign in equation (3.54) we get

$$x = b \sqrt{a - \frac{b^2}{4}},$$

(3.64)

   where we must have $|b| \leq \sqrt{2a}$.

Note that the equation with the minus sign has only one solution for x for every value of b, whereas the equation with the plus sign has two solutions.[6]

This is an example of how checking for a symmetry points to a subtle mistake in the solution of a problem. It shows the value of symmetry analysis as a tool to validate your work.

---

6. Refer to section 2.8 for an in-depth analysis of this problem using limiting cases.

3.11 Symmetric Polynomials

Section 3.1 introduced a new variable to make equation (3.3) symmetric; this helped us solve it. As we saw in subsequent examples, equations may already have a symmetry. If we decide to introduce new variables for a problem, it is often better to do so in a way that preserves or captures any existing symmetry.

For example, consider the following system of equations that must be solved for x and y:

$$\begin{aligned} x + y &= a, \\ x^4 + y^4 &= b^4. \end{aligned} \tag{3.65}$$

Note that this system of equations is symmetric with respect to swapping x and y.

One way to proceed would be to solve the first equation for $x = a - y$ and substitute the result in the second equation. This would produce a fourth-order polynomial for y. While a general solution for such polynomials is available, it is too bulky to be practical. The resulting complexity is rooted in our original decision to express x through y (or vice versa), which obfuscates the symmetry in the original system of equations.

A better way to deal with equations (3.65) is to introduce new variables that encapsulate the original symmetry. Since algebra has two commutative operations, the simplest way to construct such variables is to associate them with addition and multiplication:

$$\begin{aligned} \sigma_1 &= x + y, \\ \sigma_2 &= xy. \end{aligned} \tag{3.66}$$

Our goal is to express both equations (3.65) through these two new variables. For the first equation in (3.65), we simply use the definition of σ_1. To transform the second equation in (3.65), we do the following.

First, we square the first equation of (3.66) to get

$$\sigma_1^2 = x^2 + y^2 + 2\sigma_2. \tag{3.67}$$

Next, we isolate $x^2 + y^2$:

$$x^2 + y^2 = \sigma_1^2 - 2\sigma_2. \tag{3.68}$$

We square this equation to get

$$x^4 + y^4 + 2\sigma_2^2 = (\sigma_1^2 - 2\sigma_2)^2. \tag{3.69}$$

After expanding the right-hand side and collecting the terms, we are able to express $x^4 + y^4$ through the new variables:

$$x^4 + y^4 = \sigma_1^4 - 4\sigma_1^2\sigma_2 + 2\sigma_2^2. \tag{3.70}$$

Now the original equations (3.65) can be written as follows:

$$\begin{aligned} \sigma_1 &= a, \\ \sigma_1^4 - 4\sigma_1^2\sigma_2 + 2\sigma_2^2 &= b^4. \end{aligned} \tag{3.71}$$

We substitute σ_1 from the first equation in the second one, collect the terms, and get a quadratic equation for σ_2:

$$2\sigma_2^2 - 4a^2\sigma_2 + a^4 - b^4 = 0. \tag{3.72}$$

Its two solutions are given by the quadratic formula:

$$\sigma_{2,\pm} = \frac{4a^2 \pm \sqrt{16a^4 - 8(a^4 - b^4)}}{4}. \tag{3.73}$$

This simplifies to

$$\sigma_{2,\pm} = \frac{2a^2 \pm \sqrt{2(a^4 + b^4)}}{2}. \tag{3.74}$$

Now we recall the definitions of σ_1 and σ_2:

$$\begin{aligned} x + y &= a, \\ xy &= \frac{2a^2 \pm \sqrt{2(a^4 + b^4)}}{2}. \end{aligned} \tag{3.75}$$

We have successfully removed the fourth power in the original equations. The new system of equations for x and y is much simpler. For brevity, we denote:

$$c = \frac{2a^2 \pm \sqrt{2(a^4 + b^4)}}{2} \tag{3.76}$$

to get

$$\begin{aligned} x + y &= a, \\ xy &= c. \end{aligned} \tag{3.77}$$

If we express x through y using one of these equations and substitute the result in the second one,[7] we will arrive at another quadratic equation:

---

7. For this simpler system of equations, using an asymmetric expression for x is no longer a stumbling block.

$$x = \frac{c}{y},$$
$$y^2 - ay + c = 0, \tag{3.78}$$

where c is defined via a and b above. Solving these two equations is straightforward.

The trick of introducing variables σ_1 and σ_2 may seem artificial, but it delivers an important recommendation: try to use variables and notations that preserve the existing symmetries in your equations.

3.12 Symmetry in the Quadratic Equation

Here we revisit the ubiquitous quadratic equation

$$ax^2 + bx + c = 0. \tag{3.79}$$

Its two solutions are given by the standard quadratic formula:

$$x_{1,2} = \frac{-b \pm \sqrt{b^2 - 4ac}}{2a}. \tag{3.80}$$

Let us review the two limiting cases that we investigated in section 2.4. The first limiting case dealt with small values of $|c|$, and the second one dealt with small values of $|a|$. In both cases, one root of the quadratic equation remained finite (assuming "benign" values of the two remaining coefficients), while the other root was approaching correspondingly zero or infinity. This section shows that these two limiting cases are two sides of the same coin.

Our analysis starts from changing the variable in the original equation (3.79), assuming $x \neq 0$:

$$x = \frac{1}{y}. \tag{3.81}$$

Substitution in equation (3.79) produces the following:

$$a\frac{1}{y^2} + b\frac{1}{y} + c = 0. \tag{3.82}$$

If we multiply this equation by y^2, we will get

$$cy^2 + by + a = 0. \tag{3.83}$$

Compare equations (3.79) and (3.83). They differ only by the order of the coefficients, which is flipped. Therefore, the original equation for x is the same as the equation for $y = 1/x$, with the coefficients a and c swapped. (This is true if $x \neq 0$ and $y \neq 0$.)

This indicates that the two limiting cases we studied in section 2.4 are, in fact, closely related. The correspondence between them is summarized in table 3.1, which lists what happens if either the leading coefficient or the free term approaches zero. When interpreting the contents of table 3.1, please keep in mind that we defined $y = 1/x$, $x = 1/y$.

Table 3.1

Symmetry in the quadratic equation

Equation		Limiting case for the leading coefficient		Limiting case for the free term	
$ax^2 + bx + c = 0$	(3.79)	$x_1 \to -\frac{c}{b}$	$x_2 \to \infty$	$x_1' \to -\frac{b}{a}$	$x_2' \to 0$
$cy^2 + by + a = 0$	(3.83)	$y_1 \to -\frac{a}{b}$	$y_2 \to \infty$	$y_1' \to -\frac{b}{c}$	$y_2' \to 0$

This sets an equivalence between the pairs of cells in the table:

1. For $c \to 0$ (see the limiting case for the free term in the first row of the table), the root that is close to zero in the original equation (3.79) maps to the root that goes to infinity (limiting case for the leading coefficient in the second row) for the reciprocal of x in equation (3.83):

$$(x_2' \to 0) \Longleftrightarrow (y_2 \to \infty). \tag{3.84}$$

Similarly, for $a \to 0$ (limiting case for the leading coefficient in the first row) in (3.79), the root maps to the limiting case for the free term in equation (3.83) (see the second row):

$$(x_2 \to \infty) \Longleftrightarrow (y_2' \to 0). \tag{3.85}$$

2. For $c \to 0$, the finite root in the original equation (3.79) maps to the finite root in equation (3.83):

$$\left(x_1' \to -\frac{b}{a}\right) \Longleftrightarrow \left(y_1 \to -\frac{a}{b}\right). \tag{3.86}$$

Similarly, for $a \to 0$ in equation (3.79)

$$\left(x_1 \to -\frac{c}{b}\right) \Longleftrightarrow \left(y_1' \to -\frac{b}{c}\right). \tag{3.87}$$

We do see that equations (3.84) through (3.87) show reciprocity of the roots in each of the corresponding limiting cases $c \to 0$ and $a \to 0$. Since this symmetry stems from our original change of variables $x = 1/y$, it must also be valid beyond these limiting cases and must hold for the solutions of the quadratic equation in general. We can confirm that by using the quadratic formula. Specifically, we expect that a solution of the quadratic equation with coefficients $a, b,$ and c must be equal to the reciprocal of a solution of a quadratic equation with coefficients $c, b,$ and a. The first solution is given by the standard quadratic formula (3.80). The second solution is given by the same formula, where we swap parameters a and c:

$$y_{1,2} = \frac{-b \pm \sqrt{b^2 - 4ca}}{2c}. \tag{3.88}$$

If $x_{1,2}$ and $y_{1,2}$ are reciprocal, their pairwise products must be equal to 1. We can check this by a direct substitution. Note that this computation presents two options:

$$\begin{aligned} x_1 \cdot y_1 &= 1, \\ x_2 \cdot y_2 &= 1, \end{aligned} \tag{3.89}$$

or

$$\begin{aligned} x_1 \cdot y_2 &= 1, \\ x_2 \cdot y_1 &= 1. \end{aligned} \tag{3.90}$$

Symmetry suggests that either equations (3.89) or equations (3.90) are true, but it cannot positively identify the correct pair. Therefore, we need to check both options. Turns out that symmetry yields equations (3.90):

$$\begin{aligned} x_1 \cdot y_2 &= \frac{-b + \sqrt{b^2 - 4ac}}{2a} \cdot \frac{-b - \sqrt{b^2 - 4ac}}{2c} \\ &= \frac{b^2 - (b^2 - 4ac)}{4ac} \\ &= 1. \end{aligned} \tag{3.91}$$

We see that the symmetry here does not produce equations (3.89). This may not be very intuitive because we may expect the symmetry to "keep the numbering" of the roots, which does not happen here.

This shows that even though symmetry always finds its way from the formulation of a problem into the solution, it may manifest in a somewhat unexpected way. When performing analysis, we should not expect things to fall into our preconceived patterns but must consider different options and see what happens.

Key Point

Symmetry finds a way into equations, even if that way is not immediately intuitive.

Equations (3.90) help answer a question posed in section 2.4: what is the approximate value of the finite root of a quadratic equation when the coefficient for x^2 is close (but not equal) to zero? Obtaining an approximate solution for x when $a \neq 0$ and $a \rightarrow 0$ directly from the quadratic formula requires knowledge of calculus, which is beyond the scope of this book. Symmetry allows us to address this problem without using calculus. It relies on the fact that case $a \to 0$ is equivalent to case $c \to 0$ for the reciprocal unknown $y = 1/x$. From this reciprocity we conclude that if $a \neq 0$ and $a \to 0$, one of the roots is close to $-c/b$.

Finally, we note that if $a = c$, flipping the order of the coefficients does not alter quadratic equation (3.79). In this case, the symmetry that requires a simultaneous replacement of x by the reciprocal *and* flipping the order of the coefficients is reduced to just the first requirement. This can happen only if the original roots are reciprocal, possibly in a flipped order. We have already observed this symmetry in section 3.7. Indeed, equation (3.42) is solved by multiplying that equation by x, which does produce a quadratic equation for x, where $a = c = 1$.

We see that even the ubiquitous and seemingly trivial quadratic equation exhibits a wealth of symmetry properties. The same is true for many applications; in each case, it is our job to uncover the symmetry properties for analysis and discovery.

3.13   Linear Regression

The linear regression algorithm is widely used to find the best fit for experimental data. Section A.33 presents a basic version of this algorithm that deals with N measurements for variables x and y. These variables are not independent, and we assume there is an underlying linear relationship between them. Values y_i are affected by measurement errors, and this causes them to lie off the exact linear relationship between y and x (see figure 3.9A). Mathematically, the link between x and y is described by

$$y = ax + b + R, \tag{3.92}$$

where R is a random variable[8] that originates from measurement errors or other factors.

For example, a researcher may measure the height of plant saplings as the function of their age. The height of individual plants may vary, but in general plants grow linearly over time. To express this linear trend mathematically, the researcher may collect data for multiple plants over time. He will use x for the time since planting and y for the height of the plants in equation (3.92). Then he may use the linear regression algorithm to estimate model parameters a and b:

---

8. See section 7.1 for an explanation of a random variable.

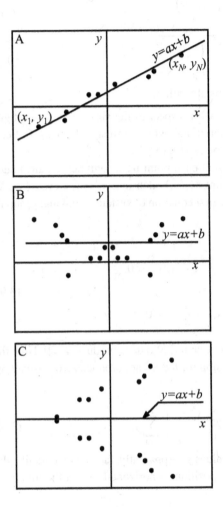

Figure 3.9
Linear regression: symmetry

$$a = \frac{N\sum_{i=1}^{N} x_i y_i - \sum_{i=1}^{N} x_i \cdot \sum_{i=1}^{N} y_i}{N\sum_{i=1}^{N} x_i^2 - \left(\sum_{i=1}^{N} x_i\right)^2},$$

$$b = \frac{\sum_{i=1}^{N} y_i \cdot \sum_{i=1}^{N} x_i^2 - \sum_{i=1}^{N} x_i \cdot \sum_{i=1}^{N} x_i y_i}{N\sum_{i=1}^{N} x_i^2 - \left(\sum_{i=1}^{N} x_i\right)^2},$$

(3.93)

where x_i, y_i are available data for variables x and y.

Here we consider several symmetries for this algorithm:

1. What happens if the data are symmetric with respect to the vertical axis (see figure 3.9B)? We expect the best-fit line to be horizontal, which corresponds to $a = 0$. Let us check this condition in the first equation of system (3.93).

   Having a symmetric data set means that each point has a symmetric pair; that is, for any point (x_i, y_i) there is another point (x_j, y_j) such that $x_i = -x_j, y_i = y_j$. We consider the sums in the numerator of the first equation of system (3.93) and group the terms by the pairs:

$$\sum_{i=1}^{N} x_i y_i = \sum_{i=1}^{N/2} (x_i y_i + (-x_i) y_i) = 0,$$

$$\sum_{i=1}^{N} x_i = \sum_{i=1}^{N/2} (x_i + (-x_i)) = 0.$$

(3.94)

Using these values in the first equation of system (3.93) does produce $a = 0$. Note that an analogous simplification of the expression for b does not produce zero. Instead, we get

$$b = \frac{\sum_{i=1}^{N} y_i \cdot \sum_{i=1}^{N} x_i^2}{N\sum_{i=1}^{N} x_i^2} = \frac{1}{N}\sum_{i=1}^{N} y_i.$$

(3.95)

We see that b is equal to the mean of all values y_i. Apparently, faced with data like that in figure 3.9B, the best-fit algorithm just positions a horizontal line in the middle of all y_i values.

2. Now we consider the case when the data points are symmetric with respect to the x axis (see figure 3.9C). From the figure, we would expect the best-fit line to coincide with the horizontal axis, which means that we should get $a = 0, b = 0$.

   We split the data points in pairs again, but in this case the pairs will have $x_i = x_j, y_i = -y_j$. Let us consider the following sums in the numerators in equations (3.93):

$$\sum_{i=1}^{N} x_i y_i = \sum_{i=1}^{N/2} (x_i y_i + x_i(-y_i)) = 0,$$

$$\sum_{i=1}^{N} y_i = \sum_{i=1}^{N/2} (y_i + (-y_i)) = 0. \tag{3.96}$$

If we substitute these values back in equations (3.93), we will see that both a and b are equal to zero, as expected.

3. Similar to other problems in this chapter, we can investigate what happens if we flip the sign of a variable. For $x_i' \leftrightarrow -x_i$, we expect to flip the slope of the best-fit line ($a' \leftrightarrow -a$), but to leave the offset intact ($b' = b$). Indeed, switching the sign of x_i in equations (3.93) yields

$$a' = \frac{N \sum_{i=1}^{N}(-x_i y_i) - \sum_{i=1}^{N}(-x_i) \cdot \sum_{i=1}^{N} y_i}{N \sum_{i=1}^{N}(-x_i)^2 - \left(\sum_{i=1}^{N}(-x_i)\right)^2} = -a \tag{3.97}$$

and

$$b' = \frac{\sum_{i=1}^{N} y_i \cdot \sum_{i=1}^{N}(-x_i)^2 - \sum_{i=1}^{N}(-x_i) \cdot \sum_{i=1}^{N}(-x_i y_i)}{N \sum_{i=1}^{N}(-x_i)^2 - \left(\sum_{i=1}^{N}(-x_i)\right)^2} = b. \tag{3.98}$$

4. Flipping the sign of y_i should change the signs of both a and b. Indeed,

$$a'' = \frac{N \sum_{i=1}^{N}(-x_i y_i) - \sum_{i=1}^{N} x_i \cdot \sum_{i=1}^{N}(-y_i)}{N \sum_{i=1}^{N} x_i^2 - \left(\sum_{i=1}^{N} x_i\right)^2} = -a,$$

$$b'' = \frac{\sum_{i=1}^{N}(-y_i) \cdot \sum_{i=1}^{N} x_i^2 - \sum_{i=1}^{N} x_i \cdot \sum_{i=1}^{N}(-x_i y_i)}{N \sum_{i=1}^{N} x_i^2 - \left(\sum_{i=1}^{N} x_i\right)^2} = -b. \tag{3.99}$$

We see that linear regression equations obey various intuitively clear symmetries that we would expect to observe in any data fit algorithm. Two more symmetries are the subject of exercise 31 at the end of this chapter. The best algorithms are those that make intuitive sense. Linear regression results are reasonable; this no doubt contributes to the popularity of this algorithm.

3.14 Summary

As we learn to recognize symmetry and invariance in mathematical problems, we gain the ability to trace them throughout a solution. It is a recurring topic of this book: the special features of a problem become enablers for a deeper analysis and for getting more robust results.

When you face a problem to solve, ask yourself if it is invariant with respect to some operation, such as swapping two variables or changing a sign of a parameter. Just as with unit consistency and limiting cases, symmetry and invariance can be used to check every equation throughout your solution and to flag an erroneous result. Yet, symmetry goes beyond the utilitarian goals of the methods we have studied above. Symmetry is always fundamental and is intimately linked with the laws of nature. It infuses a problem with meaning and often points to new knowledge.

Exercises

Problems with an asterisk present an extra challenge.

1. By swapping a and b, select three correct formulas from the six options below:

    a) $a^3 - b^3 = (a + b)(a^2 + ab + b^2)$

    b) $a^3 - b^3 = (a - b)(a^2 + ab + b^2)$

    c) $a^2 - b^2 = (a - b)(a + b)$

    d) $a^3 + b^3 = (a + b)(a^2 - ab + b^2)$

    e) $a^3 + b^3 = (a - b)(a^2 - ab + b^2)$

    f) $a^2 + b^2 = (a - b)(a + b)$

2. Check symmetry with respect to swapping α and β in the following trigonometric identities. (Hint: Use $\sin(-x) = -\sin x$ and $\cos(-x) = \cos x$.)

    a) $\sin(\alpha + \beta) = \sin \alpha \cos \beta + \sin \beta \cos \alpha$

    b) $\sin(\alpha - \beta) = \sin \alpha \cos \beta - \sin \beta \cos \alpha$

    c) $\cos(\alpha + \beta) = \cos \alpha \cos \beta - \sin \alpha \sin \beta$

    d) $\cos(\alpha - \beta) = \cos \alpha \cos \beta + \sin \alpha \sin \beta$

3. Check symmetry with respect to replacing α with $\pi/2 - \alpha$ in the following trigonometric identities. (Hint: Use $\sin(\pi/2 - \alpha) = \cos \alpha$; $\cos(\pi/2 - \alpha) = \sin \alpha$; $\sin(\pi - \alpha) = \sin \alpha$; $\cos(\pi - \alpha) = -\cos \alpha$.)

    a) $\sin 2\alpha = 2 \cos \alpha \sin \alpha$

    b) $\cos 2\alpha = \cos^2 \alpha - \sin^2 \alpha$

4. Check symmetry with respect to replacing α with $\pi - \alpha$ in the following trigonometric identities. (Hint: Use $\sin(\pi/2 - \alpha) = \cos \alpha$; $\cos(\pi/2 - \alpha) = \sin \alpha$; $\sin(\pi - \alpha) = \sin \alpha$; $\cos(\pi - \alpha) = -\cos \alpha$.)

    a) $\sin^2 \dfrac{\alpha}{2} = \dfrac{1 - \cos \alpha}{2}$

    b) $\cos^2 \dfrac{\alpha}{2} = \dfrac{1 + \cos \alpha}{2}$

5. Section A.22 considers the following equation:

    $$\frac{x - a}{x - b} + \frac{x - b}{x - a} = c. \tag{3.100}$$

    Separately, section A.23 considers a similar equation:

    $$\frac{x - a}{x - b} - \frac{x - b}{x - a} = c. \tag{3.101}$$

One of these equations is invariant with respect to swapping $a \leftrightarrow b$, and another is invariant with respect to swapping $a \leftrightarrow b$ and $c \leftrightarrow -c$ simultaneously. Using these properties, determine which of the following solutions applies to which equation:

$$x_{1,2} = \frac{(2-c)(a+b) \pm (a-b)\sqrt{c^2-4}}{2(2-c)},$$

$$x'_{1,2} = \frac{c(a+b) - 2(a-b) \pm (a-b)\sqrt{c^2+4}}{2c}, \quad (3.102)$$

where subscripts $1, 2$ correspond to the $\pm$ signs in the right-hand side.

6. Section A.15 solves the following equation for a ratio of two cosine functions:

$$\frac{\cos(\alpha+x)}{\cos(\alpha-x)} = \frac{p}{q}. \quad (3.103)$$

This problem is invariant with respect to swapping $p \leftrightarrow q$ and $x \leftrightarrow -x$. In addition, it is invariant with respect to swapping $p \leftrightarrow q$ and $\alpha \leftrightarrow -\alpha$. Use symmetry analysis to flag the wrong solutions among the following options:

a) $x = \cot^{-1}\left(\tan\alpha \cdot \dfrac{q-p}{p+q}\right) + n\pi$

b) $x = \cot^{-1}\left(\tan\alpha \cdot \dfrac{(q-p)^2}{p^2+q^2}\right) + n\pi$

c) $x = \cot^{-1}\left(\sin\alpha \cdot \dfrac{q-p}{p+q}\right) + n\pi$

d) $x = \cot^{-1}\left(\cos\alpha \cdot \dfrac{q-p}{p+q}\right) + n\pi$

7. A riverboat travels from town A to town B in time T_{AB} and from town B to town A in time T_{BA}. Section A.2 shows that the amount of time it takes to travel by a raft from town B to town A is given by

$$T_r = \frac{2T_{AB}T_{BA}}{T_{AB} - T_{BA}}. \quad (3.104)$$

A swap of towns A and B changes the sign for the time that is required to go from town B to town A on a raft. Why is the result not invariant for this swap, and what may the change of sign mean?

8. Section A.11 solves the following equation for x:

$$\frac{p}{x-a} + \frac{q}{x-b} = d. \quad (3.105)$$

The solution is as follows:

$$x_{1,2} = \frac{(p+q) + d(a+b) \pm \sqrt{d^2(a-b)^2 + (p+q)^2 + 2d(a-b)(p-q)}}{2d}, \quad (3.106)$$

where subscripts $1, 2$ correspond to the $\pm$ signs in the right-hand side. Which of the following swap symmetries are valid for this problem?

a) $a \leftrightarrow b$

b) $p \leftrightarrow q$

c) $x \leftrightarrow -x$ and $d \leftrightarrow -d$

d) $a \leftrightarrow b$ and $p \leftrightarrow q$

Check each condition in both the original equation and the solution.

9. A spherical cap is the part of a sphere that lies on one side of a plane that crosses this sphere (see section A.30 and figure 3.10). The volume V and the surface area S are given by

$$V = \frac{1}{3}\pi h^2(3R - h),$$
$$S = 2\pi Rh. \tag{3.107}$$

Note that a plane crossing a sphere creates not one but two spherical caps that are located on both sides of the plane. The sum of the volumes of these two spherical caps must equal the volume of the sphere. Similarly, the sum of the surface areas of the two spherical caps must equal the surface area of the sphere. Prove these properties by using the formulas for the volume and the surface area of a sphere:

$$V_{\text{sphere}} = \frac{4}{3}\pi R^3,$$
$$S_{\text{sphere}} = 4\pi R^2. \tag{3.108}$$

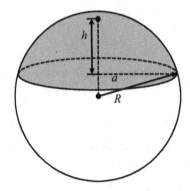

Figure 3.10
A spherical cap: symmetry

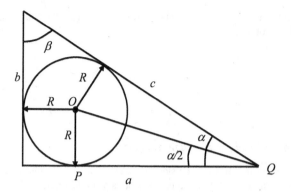

Figure 3.11
A circle inscribed in a right triangle: symmetry

10. The radius of a circle that is inscribed in a right triangle (figure 3.11) is given by

$$R = \frac{a+b-c}{2}.$$
                                                                                (3.109)

This formula is invariant with respect to only one of the following swaps of variables a, b, and c:

a) $a \leftrightarrow b$

b) $a \leftrightarrow c$

c) $b \leftrightarrow c$

Which invariance holds? Explain why.

11. Heron's formula links area S of a triangle, its half-perimeter $p = (a + b + c)/2$, and the lengths of its sides (figure 3.12):

$$S = \sqrt{p(p-a)(p-b)(p-c)}.$$
                                                                                (3.110)

Exercise 23 in chapter 2 explored a limiting case for $a \to b + c$, which leads to $S \to 0$. Use symmetry to show that a case $a \to b - c$ also results in $S \to 0$. Explain this result.

12. The number of ways to select k objects from a set of n objects is given by

$$_nC_k = \frac{n!}{k!(n-k)!},$$
                                                                                (3.111)

where a factorial for an integer is the product of all positive integers less than or equal to that integer: $n! = 1 \times 2 \times \cdots \times n$. Note that if we select k items from a set of n items, $n - k$ items are left out. Therefore, each particular selection of k objects is equivalent to a complementary selection

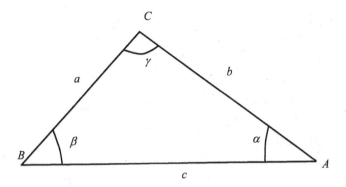

Figure 3.12
Heron's formula: symmetry

of $n-k$ objects. This creates a symmetry: the number of ways to select k objects should be equal to the number of ways to select $n - k$ objects from the same set. Prove that $_nC_k = {}_nC_{n-k}$ using the formula above.

13. Section 3.5 showed that concentrations of water in a two-syrup mix obey the same equations as concentrations of sugar. Following the same logic, show that this is also true for a three-syrup mix.

14. Explore a quadratic equation:

$$ax^2 + bx + c = 0. \tag{3.112}$$

   a) What happens to the roots if we flip the sign of the second coefficient ($b \leftrightarrow -b$)? Identify this symmetry both in the original equation and in its solution

$$x_{1,2} = \frac{-b \pm \sqrt{b^2 - 4ac}}{2a}. \tag{3.113}$$

   b) Refer to figure 2.2 in section 2.4 and sketch a plot of $x_{1,2}$ as a function of a for $b = -20; c = 10$.

15. Equations (3.78) in section 3.11 must have the same symmetry $x \leftrightarrow y$ as the original equations (3.65) in that section. Prove that this symmetry holds for equations (3.78) and then solve them for x and y and show that the symmetry holds for the solution as well. (Hint: Use the analysis in section 3.7 as a template.)

16. Consider the drawing in figure 3.13. The radii of the inscribed circle r and of the circumscribed circle R are given by

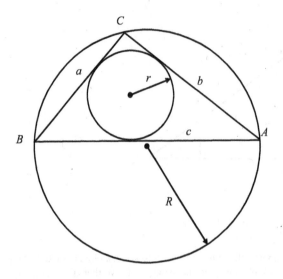

Figure 3.13
A triangle, an inscribed circle, and a circumscribed circle: symmetry

$$r = \frac{\sqrt{p(p-a)(p-b)(p-c)}}{p},$$

$$R = \frac{abc}{\sqrt{(a+b+c)(-a+b+c)(a-b+c)(a+b-c)}}, \tag{3.114}$$

where p is the half-perimeter of the triangle: $p = (a+b+c)/2$. Show that both formulas are symmetric with respect to swapping any pair of sides of the triangle.

17. Section A.9 solves the following equation for x:

$$x - \frac{1}{x} = d. \tag{3.115}$$

The solution is given by

$$x_{1,2} = \frac{d \pm \sqrt{d^2 + 4}}{2}. \tag{3.116}$$

Following the examples in section 3.7, explore the following symmetries in this problem:

a)   This equation is invariant with respect to swapping $x \leftrightarrow -\frac{1}{x}$.

b)   What happens with the roots of equation (3.115) when we flip the sign of d?

18. Exercise 13 in chapter 1 deals with pendulum oscillations. The maximum angle α_{max} between the pendulum and the vertical is called the amplitude. The formula for the period of pendulum oscillations is in the form

$$T = F(\alpha_{max}) \sqrt{\frac{l}{g}}, \qquad (3.117)$$

where $F(\alpha_{max})$ is a yet unspecified function of the amplitude, l is the length of the pendulum, and g is the gravity acceleration.

a) Do you expect the period of the pendulum oscillations to be dependent on whether the initial angle between the pendulum and the vertical was positive or negative?

b) Based on the previous question, should function $F(\alpha_{max})$ be even, odd, or neither?

19. Section A.5 solves the problem of finding intersections between a circle and a hyperbola (see figure 3.14). The circle and the hyperbola are given by the following equations:

$$x^2 + y^2 = R^2,$$
$$\frac{x^2}{H_x^2} - \frac{y^2}{H_y^2} = 1. \qquad (3.118)$$

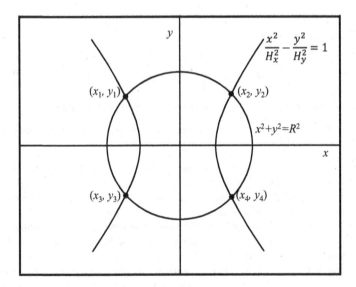

Figure 3.14
A circle and a hyperbola: symmetry

Use symmetry analysis to determine which solutions below for the intersections between the circle and the hyperbola are wrong:

a) $y = \pm \sqrt{H_y^2 \dfrac{R^2 - H_x^2}{H_x^2 + H_y^2}}$; $x_{1,2} = \sqrt{H_x^2 \dfrac{R^2 - H_y^2}{H_x^2 + H_y^2}}$; $x_{3,4} = -\sqrt{H_x^2 \dfrac{R^2 + H_y^2}{H_x^2 + H_y^2}}$.

b) $y = \pm \sqrt{H_y^2 \dfrac{R^2 + H_x^2}{H_x^2 + H_y^2}}$; $x = -\dfrac{R}{4} \pm \sqrt{H_x^2 \dfrac{R^2 + H_y^2}{H_x^2 + H_y^2}}$.

c) $y = \pm \sqrt{H_y^2 \dfrac{R^2 - H_x^2}{H_x^2 + H_y^2}}$; $x = \pm \sqrt{H_x^2 \dfrac{R^2 + H_y^2}{H_x^2 + H_y^2}}$.

d) $y = \dfrac{R}{4} \pm \sqrt{H_y^2 \dfrac{R^2 + H_x^2}{H_x^2 + H_y^2}}$; $x = \pm \sqrt{H_x^2 \dfrac{R^2 - H_y^2}{H_x^2 + H_y^2}}$.

20*. The Kalman filter computes an estimate for a quantity X that is measured by two (possibly different) instruments (see section A.32). Suppose that the first measurement produced a value X_1 with a variance of the measurement error σ_1^2, and the second measurement produced a value X_2 with a variance of the measurement error σ_2^2. Then the best estimate for X from these two measurements is given by the following equation:

$$X = \frac{X_1\sigma_2^2 + X_2\sigma_1^2}{\sigma_2^2 + \sigma_1^2}. \tag{3.119}$$

Its accuracy is characterized by its own variance:

$$\sigma^2 = \frac{\sigma_2^2 \sigma_1^2}{\sigma_2^2 + \sigma_1^2}. \tag{3.120}$$

Read section A.32 and explore the following symmetries in the Kalman filter algorithm:

a) Show that the Kalman filter algorithm is symmetric with respect to swapping measurements 1 and 2.

b) Suppose we have three measurements for the same quantity, X_1, X_2, and X_3, and variances σ_1^2, σ_2^2, and σ_3^2, respectively. The algorithm given by equations (3.119) and (3.120) tells us how to combine any two of them. We can generalize this algorithm for combining three measurements. First, we combine measurements 1 and 2. We can view the result of this computation as a virtual "measurement" with the value X and the variance σ^2, which we can then combine with measurement 3, again using the same Kalman filter algorithm. We expect that the final result of this computation should not depend on our choice of the order in which we combined measurements.[9] For example, we should get the same result if we combine measurements 2 and 3 first and then add measurement 1. Prove that this is the case.

---

9. Note that this line of argument is similar to those in sections 3.5 and 3.6.

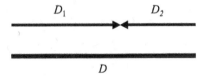

Figure 3.15
Two hikers on a trail: symmetry

21. Two hikers are starting to walk toward each other from the opposite ends of a trail (see figure 3.15). One hiker maintains speed V_1, and another maintains speed V_2. The total length of the trail is D. Section A.1 shows that the hikers will cover the following distances:

$$
\begin{aligned}
D_1 &= \frac{DV_1}{V_1 + V_2}, \\
D_2 &= \frac{DV_2}{V_1 + V_2}.
\end{aligned}
\tag{3.121}
$$

The derivation of the final result in section A.1 uses the positions $x_1(t)$, $x_2(t)$ of the two hikers as functions of time. Follow this derivation and show that the final result is invariant with respect to the following transformations:[10]

a) space shift: $x_1' = x_1 + \Delta x$ and $x_2' = x_2 + \Delta x$

b) time shift: $t' = t + \Delta t$

22. We are already familiar with Archimedes's spiral. Its radius increases linearly with the turn angle (figure 3.16). The arc length of the spiral is given by the following equation:

$$
L = \frac{\Delta R}{4\pi} \left(\theta \sqrt{1 + \theta^2} + \ln\left(\theta + \sqrt{1 + \theta^2} \right) \right),
\tag{3.122}
$$

where ΔR is the distance between the adjacent loops and θ is the total turn angle. In the figure, the spiral rolls out counterclockwise. The formula for the length should also be applicable to a

---

10. Time- and space-shift invariances are universal: all natural laws must be time- and space-shift invariant. Emmy Noether, a genius German mathematician, proved a theorem from which it follows that some invariances are intimately linked with conservation laws. Specifically, the time-shift invariance is linked with the law of conservation of energy, and the space-shift invariance is linked with the conservation of momentum.

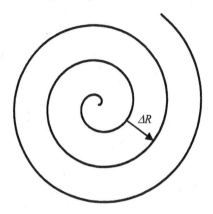

Figure 3.16
Archimedes's spiral: symmetry

spiral that rolls out clockwise. We expect that a swap $\theta \leftrightarrow -\theta$ would result either in the same value of L or in the change for the sign of L. Mathematically, this is expressed as $L(\theta) = L(-\theta)$ or $L(\theta) = -L(-\theta)$. The choice between these two options would depend on the convention for the directions for arc length computation. Prove that the second option holds: $L(\theta) + L(-\theta) = 0$.

23*. In section 3.1 we introduced a new variable to solve equation (3.3). Using that example as a template, solve the following equation for x by introducing a new variable:

$$(x - a)(x - a + 1)(x - a + 3)(x - a + 4) = b. \qquad (3.123)$$

24. Solve the following system of equations for x and y:

$$\begin{aligned} x - y &= a, \\ x^4 + y^4 &= b^4. \end{aligned} \qquad (3.124)$$

25*. Solve the following equation for z:

$$\sqrt[4]{a - z} + \sqrt[4]{z - b} = c. \qquad (3.125)$$

Use both of the following methods:

a)   Introduce new variables

$$x = \sqrt[4]{a-z},$$
$$y = \sqrt[4]{z-b}$$

(3.126)

to obtain

$$x + y = c,$$
$$x^4 + y^4 = a - b.$$

(3.127)

Then use the technique that is presented in section 3.11.

b) Shift the original variable z by introducing $v = z - z_0$. Select the value of z_0 to make the equation symmetric with respect to $v \leftrightarrow -v$:

$$\sqrt[4]{d-v} + \sqrt[4]{d+v} = c,$$

(3.128)

where d is a function of $a, b,$ and z_0. Then square and regroup the terms as necessary to produce a quadratic equation for $\sqrt[4]{d^2 - v^2}$. After that, solve for v and then for z.

26*. For what values of parameter a does the following equation have only one real solution for x (count any value only once, ignoring root multiplicity)?

$$x^{10} + e^{-a^2}x^2 - \frac{a^2 - 16}{2a} = 0.$$

(3.129)

27*. A solution for a cubic equation

$$ax^3 + bx^2 + cx + d = 0$$

(3.130)

is given in section A.29. We denote

$$\Delta_0 = b^2 - 3ac,$$
$$\Delta_1 = 2b^3 - 9abc + 27a^2d,$$
$$C = \sqrt[3]{\frac{\Delta_1 \pm \sqrt{\Delta_1^2 - 4\Delta_0^3}}{2}},$$

(3.131)

where the $\pm$ sign in this solution can be chosen arbitrarily, unless $C = 0$ for one of the signs, in which case we must choose the sign that yields a nonzero value of C. Then one of the roots is given by

$$x_1 = -\frac{1}{3a}\left(b + C + \frac{\Delta_0}{C}\right).$$

(3.132)

The solution says that the $\pm$ sign in this solution can be chosen arbitrarily (unless $C = 0$ for one of the signs). In section 3.3 we stated that an arbitrary choice in equations is associated with the symmetry with respect to that choice. Prove that cubic formula (3.132) is symmetric with respect to swapping $C_+ \leftrightarrow C_-$, where C_+ and C_- are given by the corresponding signs in the expression for C in equations (3.131). (Hint: Simplify the expression for $C_+ C_-$ and use the result to prove the desired symmetry.)

28. Section A.4 solves the problem of finding intersections between a circle and an ellipse (see figure 3.17). The circle and the ellipse are given by the following equations:

$$x^2 + y^2 = R^2,$$
$$\frac{x^2}{R_x^2} + \frac{y^2}{R_y^2} = 1, \tag{3.133}$$

where R_x, R_y are the half-axes of the ellipse. Use symmetry for this problem to flag the wrong solutions among the following options:

a) $x = \pm \sqrt{R_x^2 \dfrac{R_y^2 - R^2}{R_y^2 - R_x^2}};$ $y = \pm \sqrt{R_y^2 \dfrac{R_x^2 - R^2}{R_x^2 - R_y^2}}.$

b) $x = \pm \sqrt{R_y^2 \dfrac{R_y^2 - R^2}{R_y^2 - R_x^2}};$ $y = \pm \sqrt{R_y^2 \dfrac{R_x^2 - R^2}{R_x^2 - R_y^2}}.$

c) $x = \dfrac{R_x}{2} \pm \sqrt{R_x^2 \dfrac{R_y^2 - R^2}{R_y^2 - R_x^2}};$ $y = \pm \sqrt{R_y^2 \dfrac{R_x^2 - R^2}{R_x^2 - R_y^2}}.$

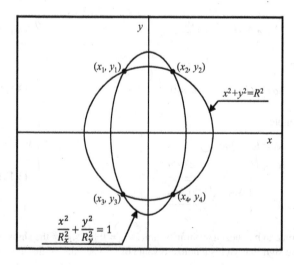

Figure 3.17
A circle and an ellipse: symmetry

d) $\quad x = \pm \sqrt{R_x^2 \dfrac{R_y^2 - R^2}{R_y^2 - R_x^2}}; \qquad y = \dfrac{R_y}{2} \pm \sqrt{R_y^2 \dfrac{R_x^2 - R^2}{R_x^2 - R_y^2}}.$

29°. The depressed cubic equation is given by

$$x^3 + px + q = 0. \tag{3.134}$$

Its three solutions can be expressed through trigonometric functions (see section A.29):

$$x_k = 2\sqrt{-\frac{p}{3}} \cos\left(\frac{1}{3} \cos^{-1}\left(\frac{3q}{2p} \sqrt{-\frac{3}{p}} \right) - \frac{2\pi k}{3} \right), \tag{3.135}$$

where $k = 0, 1, 2$. Note that equation (3.134) is invariant with respect to a change in the signs of both x and q: $x \leftrightarrow -x$; $q \leftrightarrow -q$. Prove that this invariance also holds for solution (3.135).

30. A radar measures range (distance) to the object it is tracking. Assume that there are two radars that detect a sea vessel at ranges R_1 and R_2 (see figure 3.18). Given coordinates of the radars $x_1 = 0$; $y_1 = 0$ and $x_2 = D$; $y_2 = 0$, the coordinates of the detected vessel are (see section A.25)

$$x = \frac{D^2 + R_1^2 - R_2^2}{2D},$$

$$y = \pm \sqrt{R_1^2 - x^2}. \tag{3.136}$$

Show that this solution possesses the following two symmetries:

a)   The solution is invariant with respect to $y \leftrightarrow -y$.

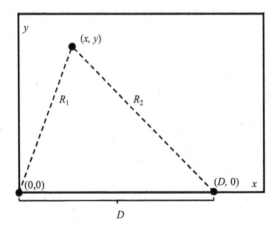

Figure 3.18
Detecting a vessel by two radars: symmetry

b)   The solution is invariant with respect to $R_1 \leftrightarrow R_2$ and $x \leftrightarrow D - x$. (Compare this with limiting cases in exercises 9a and 9b in chapter 2.)

Explain the results.

31*. Given data points x_i and y_i for two variables x and y, the linear regression algorithm estimates parameters a and b of the best-fit model $y = ax + b$ to the data:

$$a = \frac{N \sum_{i=1}^{N} x_i y_i - \sum_{i=1}^{N} x_i \cdot \sum_{i=1}^{N} y_i}{N \sum_{i=1}^{N} x_i^2 - \left(\sum_{i=1}^{N} x_i\right)^2},$$

$$b = \frac{\sum_{i=1}^{N} y_i \cdot \sum_{i=1}^{N} x_i^2 - \sum_{i=1}^{N} x_i \cdot \sum_{i=1}^{N} x_i y_i}{N \sum_{i=1}^{N} x_i^2 - \left(\sum_{i=1}^{N} x_i\right)^2}. \qquad (3.137)$$

In section 3.13 above we already considered several symmetries for the linear regression algorithm. In addition, this algorithm has two shift symmetries that are the subject of this exercise:

a)   Substitute shifted values $x_i' = x_i + c, y_i' = y_i$ in equations (3.137) and prove that they will produce new parameter estimates $a' = a$ and $b' = b - ac$.

b)   Similarly, prove that using $x_i'' = x_i, y_i'' = y_i + d$ will produce new parameter estimates $a'' = a$ and $b'' = b + d$.

c)   Explain both symmetries by considering the effects of shifts on figure 3.9.

4 Scaling

Claude Shannon was arguably one of the greatest scientists of all time. When building what is now called information theory, he introduced the definition of the bit as a unit for information (with a reference to some earlier work by John Tukey and Ralph Hartley). The very selection of this unit was partially based on the following argument.

Consider a set of relays, where each relay can be in one of two possible states. (For example, today we may refer to a two-state semiconductor cell in a solid-state storage device.) The overall state of the set carries some amount of information. The total number of combinations for N relays is doubled if another, $(N+1)$th, relay is added.[1] Yet, we would expect our measure of the information to increase by some fixed additive amount with each new relay. Shannon suggested that this can be achieved if the information is proportional to the logarithm of the total number of states. For a logarithm, doubling the input increases the value of the function by a fixed amount. Mathematically, this is expressed as $I(N) \propto \ln N$, where I is the amount of information, N is the total number of states in a system, and $\propto$ denotes proportionality. A logarithmic definition for the amount of information became one of the building blocks for developing information theory. Today, when we browse the web, the software on our device is hinged on the work that was started by Claude Shannon.

Shannon's analysis explored what happens if a variable is multiplied by a constant (the number of states is doubled in the example for a set of relays). This type of analysis is called *scaling* and is the subject of this chapter. Scaling is not unlike the symmetry introduced in chapter 3: instead of a swap or some other discrete transformation of variables, here we look at what happens if a variable x is replaced by ax, where a is a multiplier. These two topics can be viewed from a unified perspective. Partitioning them in two chapters in this book is for convenience of narration.

By far the most common case of scaling applies to problems that are formulated for quantities that are measured in certain units. The reason for this is simple: a different choice of units must leave the solution valid. Yet, a new unit of measurement is often a scaled

---

1. Indeed, for each of the two states of the new relay, we have all of the states of the original set of N relays.

version of the previous unit. For example, consider the problem about two hikers on a trail (section A.1). Two hikers walk toward each other from the opposite ends of a trail. One hiker maintains speed V_1, and another maintains speed V_2. The total length of the trail is D. Then hikers will cover the following distances:

$$D_1 = \frac{DV_1}{V_1 + V_2},$$
$$D_2 = \frac{DV_2}{V_1 + V_2}. \tag{4.1}$$

Think of what would happen if we measured distances in various units, such as miles, meters, feet, inches, parsecs, or angstroms. These different units are related to one another: in every pair they are *scaled* in a known way. Below are the scaling factors for conversion from different units to meters:

- 1 mile = 1609.344 meters
- 1 foot = 0.3048 meters
- 1 inch = 0.0254 meters
- 1 parsec $\approx 31 \cdot 10^{15}$ meters
- 1 angstrom = 10^{-10} meters

Suppose that $D_{1,2}$ are measured in meters (m), and $V_{1,2}$ are measured in meters per second (m/s). For the other units, we expect to see the same functional form of the final result, with the numerical values of all the distances and speeds scaled accordingly. Therefore, if we replace $D_{1,2}$ and $V_{1,2}$ by scaled values $D'_{1,2} = aD_{1,2}$; $V'_{1,2} = aV_{1,2}$, the a multiplier cancels and the solution remains intact. In fact, this scaling argument is the very reason behind the unit consistency rules presented at the beginning of chapter 1.

Unit consistency does not exhaust all occurrences and applications of scaling. Indeed, for the problem about two hikers, unit consistency does point to a scaling invariance, but that invariance is valid beyond a change of units. A change of units, while altering numerical values, does not affect the actual distance between the two starting points or the speeds of the hikers. Yet, if we scale actual distances and speeds by the same factor (instead of just changing the units), the solution will also remain valid. For example, the two hikers may run toward each other from the opposite ends of a football field or walk from the opposite ends of a basketball court, with proportionally lower speeds, and the former setup can be scaled to match the latter one.

The problem about the two hikers is an example of *linear scaling*: all relevant quantities are multiplied by the same scaling factor. In addition to linear scaling, we will look at other cases, such as *power law* scaling. For example, if both the height and the width of a plane figure are scaled by a factor of a, its area is scaled by a^2. For a volume of a 3D shape, we expect to have a cubic scaling.[2] Another common scaling type is *exponential*.

In this chapter, we will go through various ways that scaling can be used to solve and investigate problems.

4.1 Allometric Scaling

Some of the most striking examples of scaling apply to weight, size, and other properties of living creatures. This type of scaling is called *allometric*.

Biology offers ample opportunities to study scaling: the sizes of animals and plants vary by many orders of magnitude, and scaling of various parameters versus size becomes readily apparent. One example of allometric scaling that you can see every day at your local park: larger birds flap their wings at a lower frequency than smaller ones.

Historically, the first hypothesis about allometric scaling was formulated by Galileo. He speculated that the strength of a bone is proportional to the bone's cross-section area, which scales as the square of the size. At the same time, this strength must support the weight of the animal, which scales with its volume—that is, as the cube of the size. Therefore, bones of larger animals must be disproportionately thicker to support the weight.

Galileo's argument, while not fully accurate, is qualitatively correct. For example, smaller animals are less prone to break bones when falling from a height because the strength of their bones is better suited to withstand the consequences of the fall. A mouse can fall from 10 feet without any harm, but a horse would suffer a serious injury from such a fall.

Another example of allometric scaling is an empirical relationship called Klieber's law. It states that, for a wide variety of animals, the basal (minimal) metabolic rate B (defined as the energy burned per unit time at rest) and the mass of the animal M are approximately scaled as

$$B \propto M^{\frac{3}{4}}, \tag{4.2}$$

where $\propto$ denotes proportionality. Note that the energy metabolized by an animal is *not* proportional to its mass; the scaling here is slower. The heat from the animal's metabolism must be dissipated or the internal body tissues will overheat. Since the surface area of an animal scales as the square of its size, and the mass scales as the cube of its size, energy dissipation through the surface becomes a problem for larger animals. Arterial and bronchial networks help, but at a cost: pumping blood through the blood vessels and air through the bronchi requires spending additional energy. To make the energy balance work, larger animals have to produce less heat per unit of mass. Equation (4.2) reflects a delicate

---

2. You may notice that the scaling exponent here is equal to the number of spatial dimensions in the object. Curiously, there are geometrical shapes (called fractals) whose total length, area, or volume scales as a noninteger power of the size. In a sense, fractals have noninteger dimensionality.

balance between the need to dissipate the heat and the need to expend less energy to do that.

4.2   The Hierarchy of Scaling Behaviors

When we investigate scaling, we should keep in mind that in many applications it is valid only for a certain range of the scaling parameter. Mathematically, the scaling factor can vary by many orders of magnitude, but in practice it may have to be limited. For example, let us refer again to the two hikers problem. To make the problem more general, we should select one point for each hiker that uniquely defines his or her location, say, their centers of mass. However, in this case the original distance between the hikers must be large enough for them to be separated in space or the problem becomes ill-posed. Hence, if we have solved this problem for the original separation of 1,000 meters, we cannot scale it by a factor of 0.0001 to obtain a solution for the separation of 0.1 meter—the hikers' bodies would overlap in space.

This example, while trivial, points to a typical scenario: scaling is often applicable for a certain range of the scaling parameter, and we should use our best judgment to stay within that range.[3]

Moreover, a particular problem may exhibit multiple scaling behaviors that form a hierarchy. Consider the following quadratic function for positive values of x:

$$f(x) = x^2 + x. \tag{4.3}$$

If x is small (for instance, $x < 10^{-3}$), then the quadratic term x^2 is negligibly small compared to the linear term x. Then we get a simplified approximate expression for this function:

$$f(x) \approx x. \tag{4.4}$$

This function exhibits a linear scaling behavior: if we scale x by some factor a, the value of f is scaled by the same factor a.

In another case, we consider large values of x, such as $x > 10^3$. Then the quadratic term becomes dominant, and we get a different approximation for function $f(x)$:

---

3. Mark Twain illustrated the limits on linear scaling best: "In the space of one hundred and seventy-six years the Lower Mississippi has shortened itself two hundred and forty-two miles. That is an average of a trifle over one mile and a third per year. Therefore, any calm person, who is not blind or idiotic, can see that in the Old Oolitic Silurian Period, just a million years ago next November, the Lower Mississippi River was upwards of one million three hundred thousand miles long, and stuck out over the Gulf of Mexico like a fishing-rod. And by the same token any person can see that seven hundred and forty-two years from now the Lower Mississippi will be only a mile and three-quarters long, and Cairo and New Orleans will have joined their streets together, and be plodding comfortably along under a single mayor and a mutual board of aldermen. There is something fascinating about science. One gets such wholesale returns of conjecture out of such a trifling investment of fact."

$$f(x) \approx x^2. \tag{4.5}$$

Now the function exhibits a quadratic scaling behavior: if we scale x by some factor a, the value of f is scaled by a^2. We see that the quadratic scaling overpowers the linear one for large values of x.

Key Point

Scaling often allows us to identify a dominant phenomenon and neglect the rest. This is a great way to simplify a complex problem.

This example is not as artificial as it may seem. Early in chapter 1 we investigated the air drag force for an object moving in the air:

$$F = \frac{1}{2} C_D \rho A V^2, \tag{4.6}$$

where V is the speed of an object with respect to the air, ρ is the air density, A is the cross-section area of the object, and C_D is the drag coefficient that is a function of the shape of the object. There we concentrated on the phenomenon that is proportional to the *square* of the speed, which means that the air drag force scales quadratically with the speed of the body. However, this is not the full story. In fact, the full expression for the air drag force is attributed to competing linear and quadratic phenomena, a bit similar to equation (4.3). For example, for a sphere moving in a gas or a liquid, the linear air drag is given by *Stokes's law*:

$$F_l = 6\pi R \eta V, \tag{4.7}$$

where R is the radius of the sphere and η is the dynamic viscosity, a property of the medium. The quadratic drag is still given by equation (4.6), where $C_D \approx 0.47$.

Just as in equation (4.3), the linear phenomenon dominates at small speeds and the quadratic phenomenon dominates at large speeds. This is a case of two different scaling laws, which show up at different ranges of parameters. The transition between them corresponds to the values of the parameters (the speed and the size of the object, as well as the density and viscosity of the medium), where these two phenomena have roughly equal magnitudes. We see that the dominant phenomenon and the associated dominant scaling depend on the ratio of the quadratic term to the linear one. Since both terms measure the drag force, their ratio is dimensionless and is called the *Reynolds number*. It has practical significance: the Reynolds number, along with another dimensionless quantity, the Euler number, is used to determine the drag force for vehicles and vessels. It is not necessary to build a full-scale vessel to predict the drag force for it. Instead, engineers build a smaller model and put it in a wind tunnel or a water channel. The drag force scales; its value for the smaller model will predict the drag force for the full-scale vessel as long as the model and the full-scale vessel have the same Reynolds and Euler numbers. This application of scaling has provided enormous savings for designing and building ships, cars, and airplanes.

Key Point

Watch for constraints in scaling behavior. For some ranges of parameters, a particular scaling property may be superseded by a more powerful one.

In the above example, we have two competing scalings: a linear and a quadratic one. Both are examples of power law scaling. Extending the concept of multiple scalings further, we can think of a hierarchy of scaling behaviors for different functions. For large positive values of variable x, this hierarchy is summarized in table 4.1.

Table 4.1
The hierarchy of scaling behaviors

	x^{-a}	x^b	e^{cx}	e^{-dx}	$\ln(fx)$
x^{-p}	x^{-p}	x^b	e^{cx}	x^{-p}	$\ln(fx)$
x^q	x^q	x^b	e^{cx}	x^q	x^q
e^{rx}	e^{rx}	e^{rx}	e^{cx}	e^{rx}	e^{rx}
e^{-sx}	x^{-a}	x^b	e^{cx}	e^{-sx}	$\ln(fx)$
$\ln(ux)$	$\ln(ux)$	x^b	e^{cx}	$\ln(ux)$	$\ln(fx)$

All parameters in the table are assumed to be positive. We also assume that parameters used in the heading row are larger than the corresponding parameters that are used in the left column:

$$a > p, b > q, c > r, d > s, \text{ and } f > u. \tag{4.8}$$

The left column and the heading row show various types of scaling: power law (for example, x^b), exponential (e^{cx}), or logarithmic ($\ln(fx)$). The internal cells of the table show which of the two scalings dominates for large positive values of x. For example, when dealing with power law scaling, the larger power "wins": for $b > q$, the dominant scaling shows up as the entry x^b at the intersection of the row for x^q scaling and the column for x^b scaling. We saw an example of this behavior when we considered the quadratic scaling ($b = 2$) versus the linear one ($q = 1$): for large values of x, quadratic scaling is dominant.

As a rule of thumb, a growing exponent e^{cx} always beats any power law for large values of x, and the logarithm is the slowest growing function.[4]

If you have two exponential functions, the one with a larger parameter in the exponent wins. If $c > r$ (see inequalities (4.8)), the intersection of the column for e^{cx} in the table and the row for e^{rx} contains e^{cx} as the dominant scaling among the two.

Consider, for example, an employee whose wages grow by 3 percent a year ($r = 0.03$) and who invests a fixed percentage of her salary in a pension plan at work. Let us assume that the assets in the plan grow at 6 percent a year ($c = 0.06$). What is the long-term dynamics of the balance for that employee's pension plan? Without going through a derivation, here is the solution:

$$B = \frac{D}{c - r} \left(e^{ct} - e^{rt} \right),$$ (4.9)

where B is the balance of the pension plan as a function of time, D is the initial rate of contributions to the plan (measured in dollars per year), and t is the time in years. According to table 4.1, the balance will be eventually dominated by the exponent with the larger multiplier, that is, by e^{ct}. In this scenario, as the balance grows, the further increase will eventually be mostly attributed to the growth of the capital and not to continuing contributions. On the other hand, if $r > c$, the long-term behavior will continue to be exponential but at the rate of e^{rt}.

Since the logarithm is the inverse function for the exponent, we may expect logarithmic scaling to be some of the slowest. This is indeed the case. The *Tsiolkovsky rocket equation* links the velocity of a rocket engine's exhaust V_e, the total velocity gained by the rocket V, the total starting mass of the rocket including propellant m_0, and the final mass of the rocket when the propellant has been used m_f:

$$V = V_e \ln \frac{m_0}{m_f}.$$ (4.10)

To put a satellite with mass m_f in orbit, we need to achieve a certain threshold velocity V. Equation (4.10) exhibits logarithmic scaling of V for the total mass m_0 of the rocket at the launch pad. The "slow" logarithmic scaling explains why space rocket boosters must be so massive.

---

4. The exponent may beat a power law but may not run forever. For many real-life problems, exponential scaling is so fast that it becomes the reason for its own demise. An exponential growth increases the value of a variable until it hits a fundamental limitation that is intrinsic to that system. For example, the exponential growth of a bacteria colony in a Petri dish slows down when the total supply of nutrients is diminished or when the excretions of the bacteria accumulate and affect the growth.

4.3 Scaling and Polynomial Long Division

In this section, we illustrate the concept of scaling by applying it to the polynomial long division algorithm. As a reminder, polynomial long division starts from two polynomials $P_1(x)$ (called the *dividend*) and $P_2(x)$ (called the *divisor*) and produces the following equation:

$$P_1(x) = P_2(x)Q(x) + R(x), \tag{4.11}$$

where either $R = 0$ or the degree of R is lower than the degree of P_2. Polynomials $Q(x)$ and $R(x)$ are called, respectively, the *quotient* and the *remainder*.

The computation of polynomials Q and R is an iterative process that starts from the highest degrees of both the dividend and the divisor. Note that both sides of equation (4.11) are polynomials and that this equation must be true for any value of x. In particular, this means that it must be true for large values of x. According to table 4.1, at large values of x the highest power of a polynomial dominates any other term; therefore, the highest power terms in both sides of equation (4.11) must be equal. For example, consider the following two polynomials:

$$\begin{aligned} P_1 &= 5x^3 + 3x^2 + 2x + 4, \\ P_2 &= 2x^2 + 10x + 1. \end{aligned} \tag{4.12}$$

For these polynomials, the left-hand side of equation (4.11) has a cubic term that must match a cubic term in the right-hand side. If R is a polynomial of a lower degree in equation (4.11), then the cubic term in the right-hand side can be obtained only from a product of P_2 and Q. To get the multiplier 5 for this cubic term, the highest power of Q must be 1 and its coefficient must be equal to 2.5. Therefore, we get

$$Q = 2.5x + q, \tag{4.13}$$

where the value of the free term q is yet to be determined. Substitution of P_1, P_2, and this expression for Q in equation (4.11) will produce a polynomial equation where the cubic term cancels. Now we have quadratic polynomials, which again must be equal for all values of x. We can apply the same reasoning: their highest order terms must be equal to ensure the correct scaling for large values of x. This will produce a value for q. In general, this procedure for polynomials of an arbitrary order results in a successive estimation of the coefficients of Q starting from the highest order. As you can see, this algorithm can be viewed as one based on the scaling argument for power law functions.

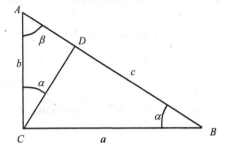

Figure 4.1
The Pythagorean theorem and a scaling argument

4.4 The Pythagorean Theorem

This section presents an ingenious argument that links scaling and the Pythagorean theorem. We start from an observation of a scaling: the area of a triangle scales as the square of its sides, as long as the scaling does not affect the angles. In other words, if we zoom in on a triangle or zoom out of it, so that lengths of the sides vary, but angles remain constant, the area scales as the square of the lengths of the sides.

For a right triangle, this can be expressed algebraically as follows:

$$S = c^2 F(\alpha), \tag{4.14}$$

where S is the area, c is the hypotenuse, and α is one of the nonright angles. We do not need to specify the exact form of function $F(\alpha)$, and it is sufficient just to state that such a function exists.[5]

With this observation in mind, we refer to figure 4.1, where we drop a perpendicular from the vertex C of the right angle to the hypotenuse. Now in addition to the original triangle ABC we have two smaller right triangles, ACD and BCD. All three triangles have the same angles $\alpha, \beta = \pi/2 - \alpha$, and $\pi/2$, and differ only by size. This means that their areas obey scaling equation (4.14):

$$S_{ABC} = c^2 F(\alpha),$$
$$S_{ACD} = b^2 F(\alpha), \tag{4.15}$$
$$S_{BCD} = a^2 F(\alpha).$$

---

5. Note that equation (4.14) is superficially similar to the correct answer for problem 13 in chapter 1 on pendulum oscillations, as it also has an unknown function of the angle as a multiplier.

Note that the equations for each triangle use their corresponding hypotenuses but the value of $F(\alpha)$ is the same for all of them.

Since triangle ABC comprises triangles ACD and BCD, the areas of the two smaller triangles sum up to the area of the larger triangle:

$$S_{ABC} = S_{ACD} + S_{BCD}. \qquad (4.16)$$

If we substitute the values from equations (4.15) in equation (4.16), the multiplier $F(\alpha)$ cancels and we get the statement of the Pythagorean theorem:

$$c^2 = b^2 + a^2. \qquad (4.17)$$

We see that scaling is intimately linked with the Pythagorean theorem! In fact, Euclid discusses scaling (without using this term) in Book VI of *The Elements* (especially proposition XXXI).

4.5 Olbers's Paradox

It is not unusual for two or more scaling properties to combine in one problem. An example of that is given by Olbers's paradox, formulated below.

First, let us consider the scaling for the intensity of light with the distance from the source. We assume that the source emits light with a constant power. Imagine two spherical shells around a source with radii R_1 and R_2. If the light is not absorbed by the medium, the amount of energy passing through the inner shell must be equal to the amount of energy passing through the outer shell (figure 4.2)—that is, must be invariant. The surface area of a spherical shell scales as R^2; therefore, the energy passing per unit area of a shell must scale as R^{-2}.

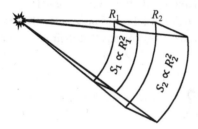

Figure 4.2
Scaling of light intensity with the distance from the source

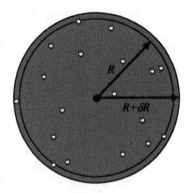

Figure 4.3
The universe in Olbers's paradox

In the formulation of Olbers's paradox this scaling is combined with another one that deals with the number of stars. Suppose that our universe is infinite and that it has an infinite number of stars. Assume that the stars are distributed with uniform average density. (Note that the stars still may be distributed nonuniformly on a local scale; only their average density on a very large spatial scale is assumed to be uniform.) Consider two closely spaced spheres centered at Earth with radii R and $R + \delta R$ (see figure 4.3). For a fixed value of δR, the volume between these spheres is proportional to the sphere area,[6] that is, scales as R^2. If the average density of the stars is uniform, then the number of stars between these spheres would also scale as R^2. Random variations may exist, but their relative magnitude would decrease for larger values of R.

As an observer on Earth, we would see the light intensity from each star scaled as R^{-2} and the number of stars at that distance scaled as R^2. These two scalings cancel out; as a result, the amount of light coming from all stars at distances ranging from R to $R + \delta R$ would scale as R^0; in other words, it will be approximately independent of R.

This creates a significant problem: if the amount of light coming from stars at distances R to $R + \delta R$ does not decrease with R, then the total amount of light coming from within the sphere of radius R_{max} on average will be proportional to R_{max}. The value of R_{max} can be made arbitrarily large, and the amount of light coming from all the stars would increase proportionally. If our universe is infinite, we should see an infinitely bright sky at night! Obviously, this does not happen, and we must seek a reason for this flawed conclusion.

Olbers's paradox can be resolved by one or more of the following factors:

1. Our universe is not infinite.

---

6. This scaling is the subject of exercise 26 at the end of this chapter.

2. It formed a finite time ago. As a result, light from distant stars has not reached Earth yet.

3. Our universe is expanding. This creates the so-called red shift, pushing light radiation from distant stars to lower frequencies and energies.

We see that an analysis of two scaling behaviors has led us to some fundamental questions about the universe we live in!

4.6 A Rope Wrapped around a Pole

To control a rope under tension, sailors wrap it around a round drum called the *capstan*. From centuries of seafaring experience, sailors know that having just a revolution or two is sufficient for a human to control a rope that is pulled by a very strong force. The reason for this is explained below.

Figure 4.4A shows what happens for some small wrapping angle α. The tension force on the left, F_1, is balanced by that on the right, F_2, and the force of friction between the rope and the capstan, F_F (technically, there is also a force F_N that is normal to the surface of the capstan, but it does not affect the calculation below):

$$F_1 = F_2 + F_F. \tag{4.18}$$

On the left-hand side the rope ties to a sail, and on the right-hand side it is pulled by a sailor. For the arrangement in figure 4.4A, all of the values in this equation are positive (we use their absolute values), and $F_1 > F_2$; this means that the sailor does not have to apply the full force F_1 to hold the rope in place. We will say that

$$F_2 = kF_1, \tag{4.19}$$

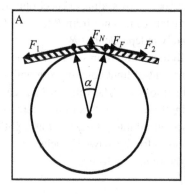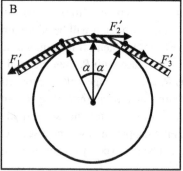

Figure 4.4
Rope tension on a capstan for a small angle

where $k < 1$. Note that k is a function of angle α: for a larger wrapping angle a sailor needs to apply a smaller force F_2 to balance force F_1.

For a fixed angle, this problem has a scaling feature: an increase in the tension force on the left would produce a proportionally larger friction force[7] and would require a proportionally larger tension force on the right to keep the rope from slipping.

We conclude that forces here obey a linear scaling: the balance between the three forces remains in effect if all forces are increased proportionally:

$$\tilde{F}_1 = \mu F_1,$$
$$\tilde{F}_2 = \mu F_2, \qquad (4.20)$$
$$\tilde{F}_F = \mu F_F,$$

where μ is an arbitrary positive scaling parameter. The key consideration here is that, for a given angle, this scaling leaves any ratio of forces constant. Therefore, for any given angle, scaling (4.20) leaves the value of k in equation (4.19) invariant.

We formulated this linear scaling for a fixed angle. Now we need to look at the force dependence on the angle, which is of greatest practical importance for the sailors. As mentioned above, k is an unknown function of the wrapping angle: $k = k(\alpha)$.

To investigate the angular dependence, let us consider what happens if the angle is doubled. Figure 4.4B shows the modified arrangement. According to our analysis above, we modify equation (4.19) for this new setup as follows:

$$F_3' = k(2\alpha)F_1'. \qquad (4.21)$$

We can split this setup in two halves, with each half being equivalent to the case depicted in figure 4.4A. Let us now look at the rope tension in the middle of the wrapping arc in figure 4.4B. If the rope tension on the left is F_1', then by virtue of the previous case (equation (4.19)) the tension in the middle is given by

$$F_2' = k(\alpha)F_1'. \qquad (4.22)$$

For the second half of the arc, we have an analogous equation:

$$F_3' = k(\alpha)F_2'. \qquad (4.23)$$

The last two equations are similar, but the corresponding magnitudes of the forces in them are different:

---

7. The maximum friction force is proportional to force F_N that is normal to the surface of the capstan and is in turn proportional to the rope tension.

$$F'_2 \neq F'_3,$$
$$F'_1 \neq F'_2. \tag{4.24}$$

Importantly, even though the magnitudes of the forces are different, the coefficient $k(\alpha)$ in equations (4.22) and (4.23) is the same, as follows from the linear scaling property (4.20). This statement is key for the following derivation.

After combining equations (4.22) and (4.23), we get

$$F'_3 = k^2(\alpha)F'_1. \tag{4.25}$$

If we compare this equation with equation (4.21), we conclude that

$$k(2\alpha) = k^2(\alpha). \tag{4.26}$$

Extending this argument for even larger multiples of angle α suggests that

$$k(N\alpha) = k^N(\alpha). \tag{4.27}$$

If we want to compute the value of k for some larger angle β, we can first estimate a number of elementary segments $N = \beta/\alpha$ and then conclude that

$$k(\beta) = k^{\frac{\beta}{\alpha}}(\alpha). \tag{4.28}$$

To make the final result look better, we introduce a new notation. We define a new variable λ as

$$\lambda = -\frac{1}{\alpha}\ln k(\alpha). \tag{4.29}$$

Note that since $k(\alpha) < 1$, we get $\lambda > 0$. Then equation (4.28) will be in the form

$$k(\beta) = e^{-\lambda\beta}. \tag{4.30}$$

Using this last expression in equation (4.19) brings us to the final result:

$$F_2 = e^{-\lambda\beta}F_1, \tag{4.31}$$

where the wrapping angle is no longer assumed to be small and λ is a positive constant. The value of that constant depends on the friction coefficient between the capstan and the rope. This problem illustrates how linear scaling for one variable leads us to the discovery of exponential scaling for another variable!

In everyday language, "exponential" is used as a synonym for "very large," and for good reason: it beats nearly any other commonly occurring function (see table 4.1 for examples). For this application, the exponential scaling is the reason that a sailor can use a turn or two around a capstan to keep a rope from sliding, even if the rope is pulled by a sail in wind that is exerting a large force. It also explains how shoelaces work: the friction force in the knot is large enough to keep the shoelaces tied.

4.7 Linear Regression

Linear regression helps researchers detect and quantify trends in data. This book presents a simple implementation that is summarized below (see also section A.33).

Suppose you have data for two quantities, x and y, and you assume that there is a dependence between these two quantities. In practice, such a dependence is never ideal because the data are always affected by measurement errors or extraneous factors. Yet, when you plot the available data points y_i versus x_i you see a linear trend. A notional example of this situation is shown in figure 4.5A. How do you quantify that trend?

We assume that y_i and x_i are linked by a linear equation

$$y = ax + b + R, \tag{4.32}$$

where R is a random variable that is responsible for the deviations from the straight line in the data (see section 7.1 for an explanation of a random variable). The linear regression algorithm states that the best estimates for parameters a and b are given by

$$a = \frac{N \sum_{i=1}^{N} x_i y_i - \sum_{i=1}^{N} x_i \cdot \sum_{i=1}^{N} y_i}{N \sum_{i=1}^{N} x_i^2 - \left(\sum_{i=1}^{N} x_i\right)^2},$$

$$b = \frac{\sum_{i=1}^{N} y_i \cdot \sum_{i=1}^{N} x_i^2 - \sum_{i=1}^{N} x_i \cdot \sum_{i=1}^{N} x_i y_i}{N \sum_{i=1}^{N} x_i^2 - \left(\sum_{i=1}^{N} x_i\right)^2}, \tag{4.33}$$

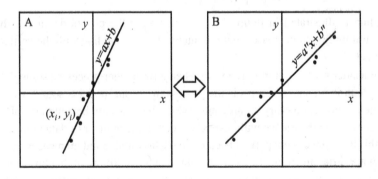

Figure 4.5
Linear regression: scaling

where x_i, y_i are the available data for variables x and y.

In the previous chapters we already discovered multiple limiting cases and symmetries in the linear regression algorithm. Here we see that it also has three scaling features:

1. Consider scaling in the vertical direction. We substitute $y_i' = \beta y_i$ in equations (4.33) to get

$$a' = \frac{N \sum_{i=1}^{N} x_i y_i' - \sum_{i=1}^{N} x_i \cdot \sum_{i=1}^{N} y_i'}{N \sum_{i=1}^{N} x_i^2 - \left(\sum_{i=1}^{N} x_i\right)^2} = \beta a,$$

$$b' = \frac{\sum_{i=1}^{N} y_i' \cdot \sum_{i=1}^{N} x_i^2 - \sum_{i=1}^{N} x_i \cdot \sum_{i=1}^{N} x_i y_i'}{N \sum_{i=1}^{N} x_i^2 - \left(\sum_{i=1}^{N} x_i\right)^2} = \beta b. \qquad (4.34)$$

This is consistent with stretching a plot of equation (4.32) in the y direction. We see that the linear regression algorithm exhibits a linear scaling for $y, a,$ and b. For example, if we calibrate our measurement instrument differently, so that it produces values βy_i instead of y_i, we expect the model for y to be scaled accordingly.

2. Next, we consider scaling in the horizontal direction. We substitute values $x_i'' = \alpha x$ in the linear regression equations to get

$$a'' = \frac{N \sum_{i=1}^{N} x_i'' y_i - \sum_{i=1}^{N} x_i'' \cdot \sum_{i=1}^{N} y_i}{N \sum_{i=1}^{N} (x_i'')^2 - \left(\sum_{i=1}^{N} x_i''\right)^2} = \frac{\alpha}{\alpha^2} a = \frac{a}{\alpha},$$

$$b'' = \frac{\sum_{i=1}^{N} y_i \cdot \sum_{i=1}^{N} (x_i'')^2 - \sum_{i=1}^{N} x_i'' \cdot \sum_{i=1}^{N} x_i'' y_i}{N \sum_{i=1}^{N} (x_i'')^2 - \left(\sum_{i=1}^{N} x_i''\right)^2} = b. \qquad (4.35)$$

This scaling is illustrated in figure 4.5. As we stretch the plot horizontally, the slope of the straight line fit decreases, but the intersection of that line with the vertical axis stays intact.

3. Finally, we consider a very common scenario when a researcher processes some data and then gets additional data and repeats the processing. If our model is sound, we do not expect the parameters to vary significantly when the new data become available. This should show as an invariance with respect to scaling the number of data points N. In practice, this invariance is not perfect because the data contain random measurement errors, which are different from one batch of data to another. To prove this invariance mathematically, we consider a hypothetical case when the measurements are fully repeatable. We compare two data sets: one with N points x_i, y_i and another with $2N$ points $\tilde{x}_i, \tilde{y}_i$, such that in the second data set all points can be grouped in pairs i and $i + 1$, where the variables are identical in each pair: $\tilde{x}_i = \tilde{x}_{i+1}, \tilde{y}_i = \tilde{y}_{i+1}$. A plot for the second data set with $2N$ points would look exactly as a plot for the first data set with N

points, but with plot markers pairwise overlaid on top of each other. For this idealized $2N$ data set we expect to get the same values of a and b. Indeed, let us consider the scaling of various sums in equations (4.33). In each sum, we group the terms by the pairs and scale the sum accordingly. All such sums will scale linearly with N. For example, we see that

$$\sum_{i=1}^{2N} \tilde{y}_i = 2 \sum_{i=1}^{N} y_i. \tag{4.36}$$

Then the expressions for a and b will scale as follows:

$$
\begin{aligned}
\tilde{a} &= \frac{2N \sum_{i=1}^{2N} \tilde{x}_i \tilde{y}_i - \sum_{i=1}^{2N} \tilde{x}_i \cdot \sum_{i=1}^{2N} \tilde{y}_i}{2N \sum_{i=1}^{2N} \tilde{x}_i^2 - \left(\sum_{i=1}^{2N} \tilde{x}_i\right)^2} \\
&= \frac{4N \sum_{i=1}^{N} x_i y_i - 4 \sum_{i=1}^{N} x_i \cdot \sum_{i=1}^{2N} y_i}{4N \sum_{i=1}^{N} x_i^2 - 4 \left(\sum_{i=1}^{N} x_i\right)^2} = a, \\
\tilde{b} &= \frac{\sum_{i=1}^{2N} \tilde{y}_i \cdot \sum_{i=1}^{2N} \tilde{x}_i^2 - \sum_{i=1}^{2N} \tilde{x}_i \cdot \sum_{i=1}^{2N} \tilde{x}_i \tilde{y}_i}{2N \sum_{i=1}^{2N} \tilde{x}_i^2 - \left(\sum_{i=1}^{2N} \tilde{x}_i\right)^2} \\
&= \frac{4 \sum_{i=1}^{N} y_i \cdot \sum_{i=1}^{N} x_i^2 - 4 \sum_{i=1}^{N} x_i \cdot \sum_{i=1}^{N} x_i y_i}{4N \sum_{i=1}^{N} x_i^2 - 4 \left(\sum_{i=1}^{N} x_i\right)^2} = b.
\end{aligned}
\tag{4.37}
$$

We see that in our thought experiment the parameters of a linear model do not change when we add more data, which we intuitively expect from a data fit algorithm.

Linear regression demonstrates predictable and reasonable scaling, in addition to sensible limiting cases and symmetries we investigated in previous chapters. When we process data for a real-life application, these intuitive properties help us "feel" the results and convey them to others.

4.8 Summary

Scaling is similar to symmetry: a transformation of a variable or variables leaves the equation intact or changes it in a predictable way. For scaling, a variable or variables are transformed using a constant multiplier; for example, x may be replaced by ax in all equations. In chapter 3, we studied *discrete* variable transformations, for example, when a variable changes a sign or two variables are swapped. For scaling, we have a *continuous* transformation where the scaling factor can have any value in some range.

Defining that range is important. For real-life applications, scaling often breaks down if the scaling multiplier is too large or too small. Moreover, a problem may exhibit multiple competing scaling behaviors, with one type of scaling replacing another at some range of the multiplier. As a result, a hierarchy of scaling behaviors may exist. For example, depending on the values of the independent variable and of the parameters, a quadratic scaling may "take over" a linear scaling or vice versa.

A scaling property in a problem helps us check a solution at each step and play through various what-if scenarios. In this sense, scaling provides us with the same benefits as limiting cases or symmetry. However, the application of scaling goes beyond that. Suppose we can solve a problem for some particular value of parameters. If the problem shows a scaling behavior, this particular solution can be scaled to apply to other parameter values. This is true for both purely theoretical problems and for practical ones. Exercise 9 at the end of this chapter shows how to leverage a previously solved problem for a new one using scaling. We have already seen how engineers can test the performance of vehicles or vessels in a wind tunnel using smaller models, as long as the parameters of the experiment ensure the correct scaling.

For yet another example, look at problem 19 in chapter 1. The speed V of a spherical shock wave from an explosion scales as a power law for the energy of the explosion E:

$$V = C \sqrt{\frac{E}{\rho R^3}}, \tag{4.38}$$

where C is a yet unknown dimensionless coefficient, ρ is the ambient air density, and R is the radius of the shock wave. As the radius R increases, the speed of the shock wave V decreases; this process can be recorded by a high-speed camera. The air density can be either measured or estimated from the weather conditions. Suppose we do just one controlled explosion with a known energy and estimate parameter C in the formula for the speed of the shock wave. Then we can use power law scaling to estimate the energy of *any* other explosion in the future by filming it from a remote location! (As you may guess, being able to do it remotely is key for this particular application.)

Exercises

Problems with an asterisk present an extra challenge.

1. There are syrups with masses m_1, m_2, and m_3 with sugar concentrations p_1, p_2, and p_3. Sections A.16 and A.17 show that blends of two or three syrups will respectively have the following concentrations of sugar:

$$p_{12} = \frac{p_1 m_1 + p_2 m_2}{m_1 + m_2},$$

$$p_{123} = \frac{p_1 m_1 + p_2 m_2 + p_3 m_3}{m_1 + m_2 + m_3}. \tag{4.39}$$

a) Is there a scaling property for the masses of syrups? What happens if we replace $m_1 \leftrightarrow am_1; m_2 \leftrightarrow am_2; m_3 \leftrightarrow am_3$?

b) Does this scaling remain valid for extremely small values of scaling multiplier a, for example, if $a = 10^{-30}$? (Hint: This is not a purely mathematical question. Think about the molecular structure of a syrup.)

c) Is there a scaling behavior for concentrations p_1, p_2, p_3?

d) Does the scaling for concentrations break down for large values of the scaling multiplier?

2. A torus is a donut-shaped body (see figure 4.6). Use scaling to flag the incorrect formulas for the volume of a torus.

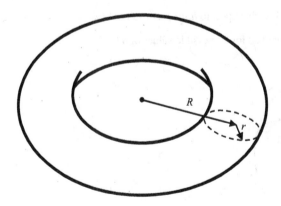

Figure 4.6
A torus: scaling

a) $V = 2\pi^2 R^2 r$

b) $V = 2\pi^2 \left(R^3 + 3R^2 r + 3Rr^2 + r^3 \right)$

c) $V = 2\pi^2 R r^2$

3. A riverboat travels from town A to town B in time T_{AB} and from town B to town A in time T_{BA}. Section A.2 shows that the amount of time required to travel by a raft from town B to town A is given by

$$T_r = \frac{2T_{AB}T_{BA}}{T_{AB} - T_{BA}}. \tag{4.40}$$

a) Is there a scaling property for travel times T_r, T_{AB}, and T_{BA}?

b) Is there a more general scaling for travel times, velocities, and the distance? Can you select such parameters α, β, and γ that replacing $T_r \leftrightarrow \alpha T_r, T_{AB} \leftrightarrow \alpha T_{AB}, T_{BA} \leftrightarrow \alpha T_{BA}, V_r \leftrightarrow \beta V_r, V_b \leftrightarrow \beta V_b, D \leftrightarrow \gamma D$ will leave the equations valid? (See section A.2 for notations.)

c) How is this scaling related to the units and dimensionality of this problem?

4. A spherical cap is the part of a sphere that lies above a plane that crosses this sphere (see section A.30 and figure 4.7). The volume V and the surface area S are given by

$$V = \frac{1}{3}\pi h^2 (3R - h),$$
$$S = 2\pi R h. \tag{4.41}$$

a) What is the scaling of S with respect to h?

b) What is the scaling of V with respect to h for small values of h?

c) Are both scalings maintained for all possible values of h?

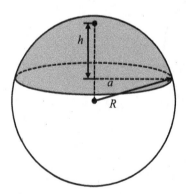

Figure 4.7
A spherical cap: scaling

5. A radar sends a powerful electromagnetic signal, which then bounces off the object that is being tracked. The radar detects the reflected signal, measures its travel delay, and estimates the distance to the object. The minimum detectable power of the received signal is an important parameter that drives the design of several radar subsystems. How does the power of the received signal scale with the following parameters of the problem:

a) The transmitted power P_t

b) The distance to the object R

c) The area of the object A_o

d) The area of the receiving antenna at the radar A_a

(Hint: See figure 4.2 in section 4.5 and the discussion there.)

6*. The Kalman filter estimates quantity X that is measured by two (possibly different) instruments. Suppose that the first measurement produced a value X_1 with a variance of the measurement error σ_1^2, and the second measurement produced a value X_2 with a variance of the measurement error σ_2^2. Then the best estimate for X from these two measurements is given by the following equation:

$$X = \frac{X_1 \sigma_2^2 + X_2 \sigma_1^2}{\sigma_2^2 + \sigma_1^2}. \tag{4.42}$$

The accuracy of X is characterized by its own variance:

$$\sigma^2 = \frac{\sigma_2^2 \sigma_1^2}{\sigma_2^2 + \sigma_1^2}. \tag{4.43}$$

Read section A.32 and explore the following scaling behaviors in the Kalman filter algorithm:

a) Values X_1, X_2 are scaled by the same factor: $\tilde{X}_1 = \alpha X_1$; $\tilde{X}_2 = \alpha X_2$.

b) Values σ_1^2, σ_2^2 are scaled by the same factor: $\tilde{\sigma}_1^2 = \beta \sigma_1^2$; $\tilde{\sigma}_2^2 = \beta \sigma_2^2$.

In what way are the implications of these two scalings for the value of X in equation (4.42) different?

7. The maximum radius of the spot on Earth's surface covered by a beam from a low-orbit satellite is given by (see section 2.12)

$$L \approx \sqrt{2RH}, \tag{4.44}$$

where R is Earth's radius and H is the orbit altitude.

a) How does L scale with the satellite orbit height H?

b) Will this scaling break down for large values of H? Why?

8. A circle is given by the following equation:

$$x^2 + y^2 = R^2. \tag{4.45}$$

Compare this with the equation for an ellipse,

$$\frac{x^2}{R_x^2} + \frac{y^2}{R_y^2} = 1,$$

(4.46)

and show that an ellipse is a scaled version of a circle.

9. A circle and a line are given by the following equations (see section A.3):

$$x^2 + y^2 = R^2,$$
$$y = px + q.$$

(4.47)

The coordinates $x_{1,2}$ of their intersections (if any) are given by

$$x_{1,2} = \frac{-pq \pm \sqrt{(1 + p^2)R^2 - q^2}}{1 + p^2}.$$

(4.48)

Use the known solution of this problem (equation (4.48)) and the results from problem 8 above to find the horizontal coordinates of the intersections between an ellipse and a straight line. Use the following equations for the ellipse and the straight line:

$$\frac{x^2}{R_x^2} + \frac{y^2}{R_y^2} = 1,$$
$$y = px + q.$$

(4.49)

(Do not solve this problem from scratch: leveraging scaling laws here is more economical!)

10. A circle and a parabola are given by the following equations:

$$x^2 + y^2 = R^2,$$
$$y = gx^2 + y_0.$$

(4.50)

Then coordinates of the intersections (if any) are given by

$$y_{1,2} = \frac{-1 \pm \sqrt{1 + 4g(gR^2 + y_0)}}{2g},$$
$$y_{3,4} = y_{1,2},$$
$$x_{1,2} = +\sqrt{R^2 - y_{1,2}^2},$$
$$x_{3,4} = -x_{1,2},$$

(4.51)

where subscripts $1, 2, 3, 4$ correspond to the $\pm$ signs in the right-hand side (see section A.6 for details). Solutions exist only if the expressions in the radicals are positive. Depending on the signs of the expressions in the radicals, there can be zero, two, or four roots.

Use solution (4.51) and the scaling from problem 8 above to find the coordinates of the intersections between an ellipse and a parabola.

11. A rectangle is inscribed in a right triangle (see figure 4.8). One of the legs has length a, and the measure of the adjacent angle is α. The side of the rectangle that is aligned with that leg has length d.

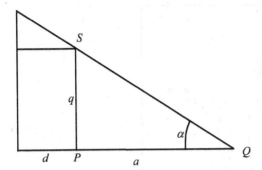

Figure 4.8
Rectangle inscribed in a right triangle: scaling

a) Use a symmetry argument to show that for an isosceles right triangle the inscribed rectangle should have an extremum (either a maximum or a minimum) for $d = a/2$.

b) Assume that the triangle and the inscribed rectangle are scaled by the same multiplier along one leg. This scaling of an isosceles right triangle generally produces a scalene right triangle. Investigate how the scaling affects the area of the inscribed rectangle and of the triangle.

c) Using the above symmetry and scaling arguments, extend the prediction of an extremum for the rectangle's area from the case of an isosceles right triangle to any right triangle. Compare this with the results in section A.28.

12. Exercise 8 in chapter 2 deals with limiting cases for the following equation:

$$\frac{1}{x-a} - \frac{1}{x-b} = d. \tag{4.52}$$

The solution for this equation for x is given in section A.12:

$$x_{1,2} = \frac{d(a+b) \pm \sqrt{d^2(a-b)^2 + 4d(a-b)}}{2d}, \tag{4.53}$$

where subscripts 1, 2 correspond to the ± signs in the right-hand side. Use a hierarchy of scalings to show that for the limiting case $|d| \to 0$ the solution produces large values for $|x|$ unless $a \approx b$. Is this also evident from equation (4.52)?

13. Section A.4 solves the problem of finding the intersections between a circle and an ellipse (see figure 4.9). The circle and the ellipse are given by the following equations:

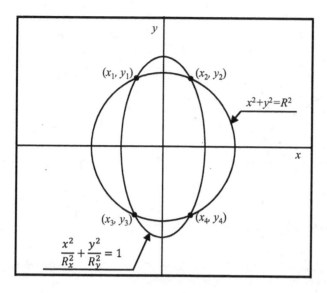

Figure 4.9
A circle and an ellipse: scaling

$$x^2 + y^2 = R^2,$$
$$\frac{x^2}{R_x^2} + \frac{y^2}{R_y^2} = 1. \tag{4.54}$$

Use scaling for this problem to flag the wrong solutions among the following options:

a) $\quad x = \pm \sqrt{R_x^4 \dfrac{R_y^2 - R^2}{R_y^2 - R_x^2}}; \qquad y = \pm \sqrt{R_y^4 \dfrac{R_x^2 - R^2}{R_x^2 - R_y^2}}$

b) $\quad x = \dfrac{1}{4} \pm \sqrt{R_x^2 \dfrac{R_y^2 - R^2}{R_y^2 - R_x^2}}; \qquad y = \dfrac{1}{4} \pm \sqrt{R_y^2 \dfrac{R_x^2 - R^2}{R_x^2 - R_y^2}}$

c) $\quad x = \pm \sqrt{R_x^2 \dfrac{R_y^2 - R^2}{R_y^2 - R_x^2}}; \qquad y = \pm \sqrt{R_y^2 \dfrac{R_x^2 - R^2}{R_x^2 - R_y^2}}$

d) $\quad x = \pm \sqrt{R_x^2 \dfrac{R_y^2 - R^{-2}}{R_y^2 - R_x^2}}; \qquad y = \pm \sqrt{R_y^2 \dfrac{R_x^2 - R^{-2}}{R_x^2 - R_y^2}}$

14. Consider a circle and a parabola that are defined by the following equations:

$$x^2 + y^2 = R^2,$$
$$y = gx^2 + y_0.$$
(4.55)

The vertical coordinates of the intersections between these two curves are given by (see section A.6):

$$y_{1,2} = \frac{-1 \pm \sqrt{1 + 4g(gR^2 + y_0)}}{2g}.$$
(4.56)

Which one of the following scaling options is valid? (Check them for both the formulation and the solution of the problem.)

a) $x \leftrightarrow ax; R \leftrightarrow aR; y \leftrightarrow ay; y_0 \leftrightarrow ay_0; g \leftrightarrow a^{-1}g$

b) $x \leftrightarrow ax; R \leftrightarrow a^{-1}R; y \leftrightarrow ay; y_0 \leftrightarrow ay_0; g \leftrightarrow ag$

c) $x \leftrightarrow ax; R \leftrightarrow aR; y \leftrightarrow a^{-1}y; y_0 \leftrightarrow a^{-1}y_0; g \leftrightarrow a^{-1}g$

d) $x \leftrightarrow a^{-1}x; R \leftrightarrow aR; y \leftrightarrow ay; y_0 \leftrightarrow ay_0; g \leftrightarrow ag$

15*. The solution of problem 6 in chapter 1 gives formulas for the velocity of gravity waves in deep water (in the case of $\lambda \ll h$) and in shallow water (in the case of $\lambda \gg h$):

$$V_{\text{deep}} = \sqrt{\frac{g\lambda}{2\pi}},$$
$$V_{\text{shallow}} = \sqrt{gh},$$
(4.57)

where λ is the wavelength (the distance between two consecutive wave crests), h is the depth of the water basin in the area of wave propagation, and $g \approx 9.8$ m/s^2 is the gravity acceleration.

a) How does the wave velocity scale with the wavelength in the deep ocean?

b) Consider shallow water waves. What happens with the wave velocity as the wave approaches a shore, where the depth is getting progressively smaller? If the depth difference causes the rear part of the wave to travel at a speed that is different from the speed of the front of the wave, what happens to the wave crest? Have you observed this effect on the beach?

c) A tsunami is an ocean wave that may be generated by an underwater earthquake. Since the area of the ocean floor that is affected by an earthquake is large, a tsunami may have a wavelength of hundreds of kilometers. Is a tsunami described by the shallow-water equation or by the deep-water equation?

d) Explain why a tsunami travels much faster than the ocean waves that you may see at the beach. (This, of course, makes a tsunami particularly dangerous.)

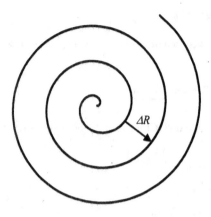

Figure 4.10
Archimedes's spiral: scaling

16. The radius of Archimedes's spiral increases linearly with the turn angle (figure 4.10). The arc
   length of the spiral is given by the following equation:

$$L = \frac{\Delta R}{4\pi} \left(\theta \sqrt{1 + \theta^2} + \ln \left(\theta + \sqrt{1 + \theta^2} \right) \right),$$ (4.58)

   where ΔR is the distance between the adjacent loops and θ is the total turn angle.

   a) What is the scaling of the arc length with respect to ΔR?

   b) Identify two scaling behaviors of the arc length with respect to angle θ that correspond to
           the two additive terms in the right-hand side of equation (4.58).

   c) Which one of these two scalings dominates the result for large values of θ?

17. The number of pairs that can be selected from N objects is given by

$$M = \frac{N(N-1)}{2}.$$ (4.59)

   a) How does M scale as a function of N for large values of N?

   b) Solve equation (4.59) for N. How does N scale as a function of M for large values of M?

18. In a room there are N people that have been randomly selected from the residents of a city to participate in a focus group. A probability p of the group having at least two people there who know each other scales approximately quadratically versus the number of people: $p(N) \propto N^2$.

    a) Does this scaling work for unlimited values of N, or does it break down for large values of N?

    b) Suppose that, for $N = 2$, the probability that these two people know each other is $p(2)$. Can you obtain an accurate result for probability $p(3)$ (that is, for the case of three people in the room) using the quadratic scaling for p? Why? (Hint: Use the results of problem 17.)

19. A chemical reaction between two different gases requires a collision between a molecule of one gas and a molecule of another gas. The speed of the reaction is directly proportional to (that is, scales linearly with) the number of such collisions per unit time.

    a) How does the speed of the reaction scale if the density of either of the gases is scaled by a factor of a, with other parameters remaining constant?

    b) Suppose we have a fixed amount of gas in a cylinder that is slowly compressed or expanded using a piston. How does the gas density scale versus its volume? (Hint: Note that gas density is measured in kg/m^3 and the volume is measured in m^3.)

    c) How does the speed of the reaction scale if the volume of a two-gas mixture is scaled by a factor b, while keeping the total amount of gas constant? (Hint: Note a similarity with problem 18 or consider the combined effects of scaling in tasks 19a and 19b for this problem.)

20. A bank promises to pay interest r on any deposit, which means that M_0 dollars in a savings account grow to $M_1 = (1 + r)M_0$ dollars in a year. This establishes a linear scaling for the amount M_1 at the end of the year versus amount M_0 in the beginning of the year. Use the derivation in section 4.6 as a template to determine how the account balance varies over time.

21. Two circles are inscribed in an angle in such a way that they touch each other (figure 4.11). Section A.26 shows that the ratio of the radii of these circles is given by

$$\frac{R}{r} = \frac{1 + \sin\frac{\beta}{2}}{1 - \sin\frac{\beta}{2}}. \tag{4.60}$$

    Equation (4.60) establishes a linear scaling for the radii of the two circles: $R \propto r$. Consider a modification of this problem where there are more than two circles inscribed in an angle. Adjacent circles touch each other. If there are N circles with radii $r_1, \ldots, r_N$, what is the ratio r_N/r_1? How does it scale with the value of N?

22. Section A.15 solves the following equation:

$$\frac{\cos(\alpha + x)}{\cos(\alpha - x)} = \frac{p}{q}. \tag{4.61}$$

    Use scaling analysis to flag the wrong solutions among the following options:

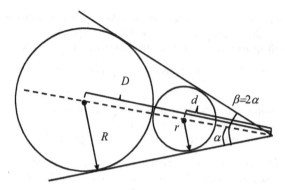

Figure 4.11
Two circles inscribed in an angle: scaling

a) $x = \tan^{-1}\left(\cot\alpha \cdot \dfrac{q-p}{p+q}\right) + pq + n\pi.$

b) $x = \tan^{-1}\left(\cot\alpha \cdot \dfrac{q-p}{p+q}\right) + n\pi.$

c) $x = \tan^{-1}\left(\cot\alpha \cdot \dfrac{pq}{p+q}\right) + n\pi.$

23. The force between two neutral atoms or molecules is commonly modeled using the *Lennard–Jones model*:

$$F(r) = \frac{24\epsilon}{\sigma}\left(2\left(\frac{\sigma}{r}\right)^{13} - \left(\frac{\sigma}{r}\right)^{7}\right), \tag{4.62}$$

where r is the distance between the atoms and σ, ϵ are positive parameters. A positive value for the force means that it is repulsive, and a negative value means it is attractive.

a) Use the hierarchy of scalings to prove that atoms attract at large distances. This is the reason that molecules in liquids and solids stay together, except at high temperatures.

b) Prove that the force becomes repulsive at small distances. This is the reason that liquids are nearly uncompressible.

24. In chemistry and biology, the Q_{10} temperature coefficient tells us that a biological process or a chemical reaction runs Q_{10} times faster if the temperature is increased by 10° C.[8]

a) What type of scaling is present for the rate of a chemical reaction with respect to the temperature?

8. Flipping this definition, a biological process or a chemical reaction runs Q_{10} times *slower* if the temperature is *decreased* by 10° C. This is why food stays fresh longer if refrigerated.

b) Suppose that two chemical reactions in a human cell have temperature coefficients Q_{10} and Q'_{10}. Assume that $Q'_{10} > Q_{10}$ and that at a normal temperature these two reactions have equal rates. Which reaction will run faster than the other one if the person has a fever?

25. If an object is freely falling in the air, its velocity increases until the force of gravity becomes balanced by the air drag force. This steady fall velocity is called the terminal velocity; for small objects, it is given by the following equation:[9]

$$mg = CR\eta V_t, \tag{4.63}$$

where m is the mass of the object, g is the gravity acceleration, C is a coefficient, η is the air viscosity, R is the size of the object, and V_t is its terminal velocity. Consider a steady fall scenario and investigate the scaling with respect to the size of the object:

a) If the density of the object is constant, what is the scaling of the object's mass with its size?

b) Knowing the scaling for mass m, determine the scaling for V_t versus the object's size that satisfies the equation above.

c) What does this scaling tell us for extremely small objects? Explain why specks of dust that you see in a ray of sunlight do not seem to fall down.

26. Section 4.5 noted that the volume of the space between two concentric spheres of radii R and $R + \delta R$ scales as R^2 (see figure 4.3).

a) Prove this quadratic scaling from the formula for the volume of a sphere (assume that $\delta R \ll R$):

$$V = \frac{4}{3}\pi R^3. \tag{4.64}$$

b) How does the volume between the two spheres scale as a function of δR if $\delta R \ll R$?

c) How does the volume between the two spheres scale as a function of δR if $\delta R \gg R$?

d) What is the condition for δR that corresponds to the transition between these two last scalings?

27. The discussion of equations (4.6) and (4.7) in section 4.2 explores competing scalings with respect to velocity for an object that is moving in the air. Another variable used in both equations is the size of the object. Consider a hailstone that starts forming in a cloud as a tiny speck and then grows to a larger ball. As it moves through the air, it is subject to the air drag force.

Which of the two mathematical models (given by equations (4.6) and (4.7)) describes the air drag force for the initial stages, and which is appropriate for the later stages of the hailstone growth? (Hint: Take into account that radius R and the cross section of the hailstone A in these two mathematical models are linked. Assume that the speed of the fall increases as the hailstone grows.)

---

9. For a sphere, $C = 6\pi$, the formula known as Stokes's law (see section 4.2).

28*. Consider a plot of the function $y = 1/x$ for positive values of x.

    a) Using scaling, prove that the area $S(a)$ under the curve $y = 1/x$ for $1 \leq x \leq a$ is the same as the area under the same curve for $c \leq x \leq ca$ for any positive value of c.

    b) Consider the value of $S(a^2)$, which is the area under the same curve for $1 \leq x \leq a^2$. Split this area in two sections: area $S(a)$ for $1 \leq x \leq a$ and area $S'(a)$ for $a < x \leq a^2$. Using the result from task 28a, express area $S(a^2)$ through $S(a)$. This establishes a scaling for function $S(a)$.

    c) Which standard function exhibits the same scaling property as function $S(a)$?

29. As workers gain more experience, they perform their tasks faster. The first quantification of this effect is known as Wright's law. In 1936, T. P. Wright observed that as the production of airplanes at the Curtiss-Wright factory doubled, the time required to manufacture each airplane decreased by 20 percent. Express Wright's law as a scaling relationship.

30*. The depressed cubic equation is given by

$$x^3 + px + q = 0. \tag{4.65}$$

Its three solutions can be expressed through trigonometric functions (see section A.29):

$$x_k = 2\sqrt{-\frac{p}{3}} \cos\left(\frac{1}{3} \cos^{-1}\left(\frac{3q}{2p}\sqrt{-\frac{3}{p}}\right) - \frac{2\pi k}{3}\right), \tag{4.66}$$

where $k = 0, 1, 2$.

    a) Show that equation (4.65) obeys the scaling property:

$$\begin{aligned} x' &= ax, \\ p' &= a^2 p, \\ q' &= a^3 q. \end{aligned} \tag{4.67}$$

    b) Show that this scaling applies also to solution (4.66).

5 Order of Magnitude Estimates

On July 16, 1945, the United States conducted an atmospheric nuclear explosion test. How could one measure the power of that explosion? The great Enrico Fermi was observing the test from a ten-mile distance, where the effect of the blast could be felt but supposedly was not dangerous. He was dropping small pieces of paper. When the blast wave from the explosion reached his location, it pushed the falling pieces of paper by about $2\frac{1}{2}$ meters. This observation alone was sufficient for Fermi to roughly estimate the power of the explosion as equivalent to about 10,000 tons of TNT. The estimate was quick and *very* approximate, but it worked.

In this chapter we will learn how to make quick and dirty estimates for various quantities. The need for such estimates is real and here is why: We have already seen that there may be two or more competing phenomena that contribute to a final result. Contrary to the impression created by some standard math problems, this is the norm and not the exception. Consider, for example, an engineer who is tasked with modeling the orbit of the next telecommunications satellite. A standard theory, going back to Newton, states that satellites move in elliptical orbits. However, a closer examination shows that this theory assumes that there are only two point or spherical masses, Earth and the satellite, and that the motion is determined only by the gravity force. In the real world these assumptions are clearly violated. The engineer compiles a list of phenomena and factors that may affect the orbit:

1. The gravity force, assuming that Earth is a uniform sphere.
2. A correction to the spherical model, caused by the geoid shape of Earth (a sphere, slightly squeezed at the poles).
3. A correction to the geoidal model, caused by the nonuniform distribution of mass in the geoid.
4. The atmospheric drag force. We are already familiar with this phenomenon for objects moving in the air from chapter 1. A satellite is not moving in a complete vacuum, even though the atmosphere at the altitude of the satellite orbit is very thin. A small drag force will be present.

5. Earth is not the only source of gravity. There is the Sun, the Moon, and other planets (Mars, Venus, Jupiter, etc.).

6. The shape of Earth's oceans is not static. There are tides in the ocean that bulge it out as the positions of the Sun and the Moon change.

7. Solid Earth tides: we are familiar with ocean tides that are caused by the gravitational force from the Sun and the Moon, but the same force also distorts the shape of the solid Earth ever so slightly.

8. A light pressure: the sunlight bounces off the satellite, pushing it slightly in the direction away from the Sun.

9. A solar wind pressure: in addition to the visible light, the Sun emits a flux of particles. We do not experience this radiation because it is stopped by the dense layers of the atmosphere below satellite altitudes, but a satellite will be pushed by it.

10. General relativity effects: the exact mathematical model for a body motion in a gravitational field is given by Einstein and is different from Newton's theory.

A full account of all these phenomena would be immensely complex and may not be feasible. Moreover, the complexity of this gargantuan task would not scale linearly with the number of factors that we account for. Different factors become coupled, further increasing the complexity. For example, while the gravitational force from Mars affects the satellite directly, it also contributes to ocean tides, which change the shape of the oceans, in turn perturbing the orbit of the satellite.

The only realistic way to complete the task of modeling the orbit is to cut a few corners. For example, the engineer may come to the conclusion that the general relativity corrections are too small to worry about, but ocean tides must be taken into account. Unfortunately, this tactic creates a catch-22 problem: to neglect a factor among those listed above, we must prove that this factor is small enough, which requires estimating it first.

In practice, this conundrum is resolved by making preliminary, rough estimates for all relevant factors. The standard for the accuracy of such estimates is very low because they serve only to convince us that some of the phenomena are too insignificant to worry about. The art of making such estimates is the subject of this chapter.

Key Point

When dealing with a complex problem, start from the most important phenomenon. Then work your way down by including smaller effects until you reach the desired accuracy of your model.

The difficulty bar for a rough estimate is low: we can overestimate or underestimate a value by a factor of two or three and such a crude estimate may still be helpful. Often this low bar drastically simplifies making estimates compared to bona fide calculations.

When armed with the skill to produce rough estimates, we can set up a plan of attack on a complex, multifaceted problem:

1. Identify phenomena that contribute to the final result.
2. Produce rough estimates for these phenomena. Keep calculations as simple as possible.
3. Make a decision on which phenomena must be retained for future analysis and which can be neglected.
4. Focus on the phenomena that cannot be neglected and ignore the rest.

The use of rough estimates is not limited to finding candidate factors to neglect in the analysis. Occasionally, we can also use them to check our results. Crude estimates can provide only a crude check of the result and thus are not always definitive. Yet, it is not uncommon to find an error in a solution using such an estimate (see section 5.3 for an example).

5.1   How Good Should an Estimate Be?

Rough estimates often target the *order of magnitude* of a quantity. A common definition for the order of magnitude of x is the value of the base 10 logarithm of x, rounded to the whole-number floor, that is, to the largest integer not exceeding the value of $\log_{10} x$. For the purpose of making rough estimates, it is more convenient to round the logarithm to the *nearest* integer and not to the floor. We will use this second criterion below.

Orders of magnitude are often used to compare two quantities. If we say that a is on the order of b, this means that they have the same orders of magnitude. We will denote this as

$$a \sim b, \tag{5.1}$$

which means that

$$\lfloor \log_{10} \frac{a}{b} \rfloor = 0, \tag{5.2}$$

where $\lfloor \rfloor$ denotes a rounding to the nearest integer.

For example, if we say that a is on the order of b where $b = 1$, this means that the value of a may roughly range from 0.3 to 3. Indeed, if the value of a is between $\frac{1}{\sqrt{10}} \approx 0.3$ and $\sqrt{10} \approx 3$, then $-0.5 < \log_{10} a < 0.5$, and therefore $\log_{10} a$ is rounded to zero, satisfying equation (5.2) for $b = 1$.

The condition for variable c to be small in comparison to variable d is denoted as

$$c \ll d, \tag{5.3}$$

which means that

$$\lfloor \log_{10} \frac{c}{d} \rfloor < 0. \tag{5.4}$$

In this definition, c is at least one order of magnitude smaller than d. As before, $\lfloor \rfloor$ denotes a rounding to the nearest integer. For example, we may observe that c is two orders of magnitude less than d. This would mean that the value of c/d may range roughly from 0.003 to 0.03. If the ratio of two numbers g and h is $g/h = 0.09$, we say that g is one order of magnitude[1] less than h.

Similarly, a quantity d being large compared to c is denoted as

$$d \gg c, \tag{5.5}$$

which means that

$$\lfloor \log_{10} \frac{d}{c} \rfloor > 0. \tag{5.6}$$

It goes without saying that any comparison of magnitudes can be done only if the variables are measured in the same units or if both are dimensionless. Then the ratio of these variables is dimensionless, and we say that one of them may be "small" and another may be "large."[2]

The above criteria for smallness and largeness are not set in stone and should be applied with a particular goal in mind. What is "small enough" for one problem may not be sufficiently small for another one. In some cases you would need two orders of magnitude difference or even more.

5.2   How to Make Order of Magnitude Estimates

Making order of magnitude estimates is more an art than a science. That is why the guidelines below should not be taken as hard rules. Still, it is helpful to keep in mind the following (which are referred to as rules throughout this chapter):

1. Often there are multiple ways to get the desired estimate. Use your common sense to select the path of least resistance.

2. Assume that different phenomena are not coupled. Neglect the coupling effects unless it is the coupling itself that you are after.

---

1. Note that if we used rounding to the floor in the definition of the order of magnitude, then a ratio $g/h = 0.09$ would mean that g is two orders of magnitude less than h. Intuitively, we expect to see one order of magnitude difference here, not two, because 0.09 is so close to 0.1, and 0.1 corresponds to exactly one order of magnitude difference.

2. We already encountered the notion of small and large values early in chapter 1, when we explored dimensionless variables.

3. Do not chase the best accuracy. Try to capture only the simplest mechanism for each phenomenon and use the simplest model that describes it.
4. For a varying parameter, use a single "typical" value where appropriate unless you are specifically estimating the variation of the parameter. The typical value can be a median value of the total range of variation or an estimate for a mean.
5. Estimate a product or a ratio of several factors as a product or a ratio of the estimates for individual factors.
6. Estimate a sum or a difference of several terms as equal to the term that has the largest absolute value.
7. If the number of terms is large ($N \gg 1$), estimate their sum as the typical value of one term times the number of the terms.
8. If you have to estimate a magnitude of variation for a parameter rather than a single value, try the following procedure:

   (a) If possible, define the magnitude of variation as the *significant change*. The significant change is estimated to be on the order of a typical or the maximum range of variation for that parameter. Often the magnitude of variation is on the order of the typical value of that parameter as described in rule 4 above.

   (b) If it is difficult to estimate the significant change for a variable f, see if this variable is a function $f(x)$ of another variable x. See if the magnitude of variation for x can be estimated. If yes, assume that the relative variation of the function is the same as the relative variation of the independent variable:

$$\frac{\Delta f(x)}{f(x)} \sim \frac{\Delta x}{x},$$ (5.7)

   where Δx is the magnitude of a significant change in x.

   (c) When estimating a relative variation of a product or of a ratio of two variables, use the sum of the relative variations for these variables:

$$\frac{\Delta(fg)}{fg} \sim \frac{\Delta f}{f} + \frac{\Delta g}{g}.$$ (5.8)

   Then apply rule 6.

   (d) When estimating a relative variation of a function of two independent variables, use the sum of the relative variations for these variables:

$$\frac{\Delta f(x, y)}{f(x, y)} \sim \frac{\Delta x}{x} + \frac{\Delta y}{y}.$$ (5.9)

   Then apply rule 6.

9. If using a formula that is obtained using dimensional analysis and that has an unknown dimensionless multiplier, assume that the value of that multiplier is on the order of one.[3]

These rules may be not clear for now. Examples below will demonstrate how they work in practice.

5.3   Mortgage Payments

This section illustrates some of the rules for making order of magnitude estimates to correctly compute mortgage payments (see section A.31 and exercise 1 in chapter 1). For the initial loan amount D, the interest rate r, and the duration of the loan T, the payment rate for the mortgage is given by

$$p \approx \frac{rDe^{rT}}{e^{rT} - 1}. \tag{5.10}$$

Jane used formula (5.10) to estimate payments on a mortgage that she is planning to apply for. She used loan amount $D = \$230,000$, the interest rate 4.5 percent ($r = 0.045$/year), and the loan duration of $T = 30$ years. Using these values, she estimated her payments (principal and interest) to be $p = \$13,972.14$, which seemed very high. The loan officer at the bank said that the correct amount of her monthly payment was $p_0 = \$1,164.35$.

To reconcile the numbers, Jane decided to come up with an order of magnitude estimate. Since p is one order of magnitude greater than p_0, an order of magnitude estimate should be able to point to the correct value.

Jane reasoned that a mortgage payment is a sum of two terms: contributing to the principal and paying interest. She decided to compute them separately, not accounting for a possible coupling (see rule 2). She started from computing the contribution to the principal and then proceeded to computing the interest portion:

1. The mortgage term of 30 years means that the loan amount must be paid off in 360 monthly payments. Therefore, the average monthly principal payment is given by

$$p_p = \frac{D}{12T}. \tag{5.11}$$

   As the mortgage is being paid off, the unpaid loan amount decreases. This means that each month Jane will be paying less for the interest and more to the principal. Thus, the payment to the principal will vary with time. For the order of magnitude estimate, Jane used a crude way to obtain her estimates (see rule 3). She ignored the time variation in p_p and assumed that the principal is paid off at a constant rate (see rule 4).

---

3. This rule may look as if it is pulled from thin air, but in practice it works with amazing regularity.

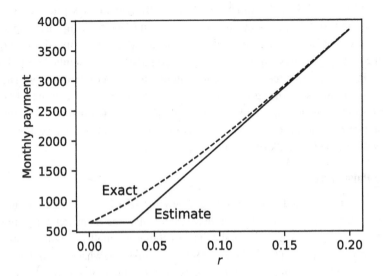

Figure 5.1

Monthly mortgage payments as a function of the interest rate

2. Next, Jane estimated the amount that must be paid for interest. On balance D, the annual interest comes up to Dr. The monthly interest payment is given by $Dr/12$. Again, Jane ignored the variation of that amount over time.

She computed the total monthly payment as the greater of the two components (see rule 6):

$$p_{est} = \max\left(\frac{Dr}{12}, \frac{D}{12T}\right). \tag{5.12}$$

Using the numerical values for this particular mortgage, she got $p_{est} =$ \$862.50. This is the same order of magnitude as the value provided by the loan officer ($p_0 =$ \$1,164.34) and is an order of magnitude lower than the value from equation (5.10). After some thinking, Jane realized that equation (5.10) computes the annual payment, and not the commonly used monthly amount. These values differ by a factor of 12, or by one order of magnitude.

Key Point

Rough estimates are a blunt but useful instrument to detect flaws in your work. They can also serve as a starting point in building a mathematical model.

Let us investigate how well Jane's estimate works for different interest rates. Figure 5.1 shows a plot of the monthly payment obtained from equation (5.10) by dividing p by 12 (the dashed line) and the order of magnitude estimate (5.12) (the solid line). The estimate has a "knee" that separates two regions: estimated payments are applied entirely to the principal for lower interest rates and to paying interest for higher interest rates. The selection of the dominant mechanism is governed by the max() function in equation (5.12). Overall, we get a decent estimate, especially if one component truly dominates. When the two components are of the same order, their effects become coupled and the rough estimate is off a bit, but it still predicts the bulk of the amount correctly.

5.4   Designing a Parachute

Mike is an engineer who is tasked with designing a new model of a parachute entirely from scratch, not just modifying an existing model. Specifically, the management of the parachute manufacturing company wants to make the parachutes as small as possible without jeopardizing safety. To approach this problem, Mike must first find what is feasible and what is not. To determine the minimum possible canopy size, he starts by making an order of magnitude estimate.

First, Mike estimates the maximum safe landing speed. He observes (see rule 1) that a trained person should not have a problem jumping from the height of 10 feet, or ~3 meters. He uses dimensional analysis[4] to conclude that the landing speed of a person jumping from height h is given by

$$V_l = C_v \sqrt{gh}, \tag{5.13}$$

where g is the gravity acceleration and C_v is an unknown dimensionless constant. (Note that in this formula, h is not the height of a jump with the parachute but a height of a safe jump without a parachute.)

The parachute must be large enough to ensure this or lower value of the landing speed, regardless of the height that the parachute user may have jumped from. This means that the person's speed must stop increasing after some initial time in the air and must remain at a safely low value for the rest of the descent. This occurs when gravity is balanced by the air drag force, that is, if the person descends at the so-called terminal velocity, given by the following equation:[5]

$$mg = \frac{1}{2} C_D \rho A V_t^2, \tag{5.14}$$

4. See problem 9 in chapter 1 for a formulation of this dimensional analysis problem.

5. Note that this equation is different from that in problem 25 in chapter 4 because the latter applies to small objects at low speeds.

where m is the mass of the person and the parachute, g is the gravity acceleration, C_D is the drag coefficient for the parachute, ρ is the air density, A is the cross-section area of the parachute, and V_t is its terminal velocity.

To ensure safety, the terminal velocity must be less than or equal to the safe landing speed: $V_t \leq V_l$. The smallest possible canopy corresponds to the maximum terminal velocity, and we expect to have $V_t = V_l$. Mike arrives at the following equation:

$$m = \frac{1}{2}C_D A \rho C_v^2 h. \tag{5.15}$$

Note that even though this equation does not contain gravity acceleration, the height of a safe jump h is still dependent on it implicitly. Solving equation (5.15) for the parachute area A yields

$$A = \frac{2m}{C_D C_v^2 \rho h}. \tag{5.16}$$

Now Mike proceeds to making a numerical estimate. He guesses (see rule 9) that C_v is on the order of 1. He knows that for a sphere the drag coefficient is approximately equal to 0.47. However, a sphere seems to be better suited to have the air flowing around it than a parachute canopy. In fact, the latter is specifically designed to have *more* air drag. The engineer decides to use $C_D = 1.5$. (By the way, this is another example of applying rule 9!)

For the weight of the person plus the weight of the parachute, 80 kg seems like a reasonable estimate. The air density is about 1.2 kg/m^3. Then the parachute area is estimated as

$$A \sim \frac{160}{1.5 \cdot 1.2 \cdot 3}\text{m}^2 \approx 30 \text{ m}^2. \tag{5.17}$$

Note that this equation contains a product and a ratio of several variables (m, C_v, C_D, h), for which we have roughly estimated their values separately. The estimate for A is obtained by plugging these separate estimates (see rule 5) in equation (5.16).

This corresponds to roughly 330 square feet (common units for this application).[6] Of course, for a real design this estimate should be refined using a better model and carefully designed experiments. Yet, it produces an initial idea of what may or may not be feasible for a brand new parachute model. This estimate shows that it may be impossible to decrease the canopy size by a large amount below 330 square feet. By looking at equation (5.16), the engineer concludes that a limited opportunity to reduce the parachute area from the current size may only come from changing the shape of the canopy to increase the air drag

---

6. Turns out, Mike has overestimated the typical canopy size by about factor of 1.5, but he is still within the correct order of magnitude.

coefficient C_D. For this problem, an order of magnitude estimate roughly defines the realm of possibilities and even suggests a way to improve the design of a product.

5.5   Accuracy of a Pendulum Clock

How accurate is a typical grandfather clock? The time measurement for such clocks is based on counting pendulum oscillations. A clock can be fast or slow if the actual period of pendulum oscillations differs from the design value.

From the answer to problem 7 in chapter 1, the period of pendulum oscillations is given by

$$T = C \sqrt{\frac{l}{g}}, \tag{5.18}$$

where C is a constant dimensionless coefficient, l is the length of the pendulum, and g is the gravity acceleration.

A clock designer assumes certain values for l and g and builds a clock for these precise specifications. In reality, the values of l and g may slightly differ from the spec, and the clock will be a bit fast or a bit slow as a result.

To estimate the typical accuracy of a pendulum clock, we need to estimate the variability of T that is caused by a variability of l and g. This is the case for estimating the variation in the dependent variable (T) that stems from the variation in two independent variables (l and g). We will consider them in turn:

1. The reason for the variability in the pendulum length l is that the manufacturing process is never absolutely accurate. If the pendulum length for a floor clock is $l = 1$ m and a manufacturing tolerance is 0.1 mm, the relative error in the pendulum length is

$$\frac{\Delta l}{l} \sim 10^{-4}. \tag{5.19}$$

   We assume this value applies to the relative error in the period of oscillations as well (see rule 8b):

$$\frac{\Delta T_l}{T} \sim 10^{-4}, \tag{5.20}$$

   where subscript l denotes a contribution from the error in the length of the pendulum.

2. The reason for the variability in the gravity acceleration is a bit more subtle. One mechanism here is that Earth is not perfectly spherical but has an oblate shape.[7] Therefore, we assume that the gravity acceleration at a high latitude (such as in Norway) may be a bit different from that at a low latitude (for instance, in Singapore). For the sim-

---

7. For this problem, we ignore the effects of Earth spinning and of the terrain. They can be estimated separately.

plest gravitational model, the force of gravity is a function of the attracting mass and the distance to it. To proceed with our order of magnitude estimate, we will use this simplest model (see rule 3).

Moreover, for the distance to the attracting body we just use the distance to the center of Earth, even though in reality the gravity acceleration is affected by a complex distribution of mass around it. Hence, the gravity acceleration is assumed to be a function of the Earth mass M and the distance R to the center of Earth: $g = g(M, R)$. The Earth mass is the same for all locations, but R depends on the latitude. The variability of g now can be estimated (again, see rule 8b) as

$$\frac{\Delta g}{g} \sim \frac{\Delta R}{R}, \tag{5.21}$$

where ΔR is the difference between the equatorial and the polar radii. The variability of g directly flows through to the variability of the period of pendulum oscillations:

$$\frac{\Delta T_g}{T} \sim \frac{\Delta g}{g} \sim \frac{\Delta R}{R}, \tag{5.22}$$

where subscript g denotes a contribution from the error in the gravity acceleration.

The equatorial radius of Earth is about $6,378$ km, and the polar radius is about $6,357$ km. Then $\Delta R/R \approx (6378 - 6357)/6378 \approx 0.0033$. We may expect the gravity acceleration to vary by the same relative amount. For the base value of $g = 9.8$ m/s^2, we get $\Delta g \sim 0.0033g \approx 0.033$ m/s^2 for the variation in the gravity acceleration.[8]

With these estimates in hand, we can compute the accuracy of a pendulum clock. For a manufacturer that ships clocks worldwide, from Norway to Singapore, the biggest danger to accuracy may come from the variability of the gravity acceleration. Fortunately, this error is predictable and easy to compensate for: they can have an adjustable screw on the pendulum that would be used to fine-tune the effective length and that can be preset differently for customers at different latitudes.

In addition to that, we saw that a random manufacturing error in the pendulum length may produce a relative error in the clock on the order of $\sim 10^{-4}$. In a day ($\sim 10^5$ s) this error may accumulate to about 10 seconds or so.

If both causes are taken into account, we should use a sum of their relative contributions to arrive at the total relative error for the clock (see rule 8c).

---

8. In fact, the gravity acceleration varies by ~ 0.07 m/s^2 depending on the location, which is about twice our estimate, but definitely within the same order of magnitude. We have made this estimate using the crudest assumptions possible, yet got it right!

5.6 Sizing the Power for a Car Engine

What should the power of a car engine be? In classical mechanics we learn that the power P needed to counter a particular force $\vec{F}$ is given by

$$P = -\vec{F} \cdot \vec{V}, \qquad (5.23)$$

where $\vec{V}$ is the velocity of a moving body and $\vec{F} \cdot \vec{V}$ is a dot product[9] of vectors $\vec{F}$ and $\vec{V}$.

To estimate the engine power required for normal car use, we first need to determine how that power is used. Primarily, it is spent on countering two forces, which we will consider in turn.

1. *Overcoming the air drag force.* We are already familiar with the air drag force (see equation (1.2) in chapter 1 and the discussion there):

$$F = \frac{1}{2} C_D \rho A V^2, \qquad (5.24)$$

   where ρ is the density of the air, A is the cross-section area of the car, and C_D is called the drag coefficient. The air drag force has the direction that is exactly opposite to the car velocity, and therefore the dot product is equal to the negative of the product of the air drag force magnitude and the car speed. This yields

$$P_a = \frac{1}{2} C_D \rho A V^3, \qquad (5.25)$$

   where subscript a denotes the portion of the engine power that is needed to overcome the air drag.

2. *Working against the force of gravity when driving uphill.* In this case, the car that climbs up must work against the downward-directed gravity force. The physics of the body motion on inclined surfaces is beyond the scope of this book, but we can make rough estimates even if we are not familiar with the exact theory. To do that, we start from two limiting cases:

   (a) For a body that moves vertically, the gravity force in the direction of the body motion is just the weight of that body, or $F_g = -mg$, where m is the mass and g is the gravity acceleration. Of course, a vertical road is a purely hypothetical case. Since the car does not move straight vertically, the contribution from the gravity force must be smaller than mg.

   (b) If a car drives on a level road (at zero incline), the engine spends no additional power on countering the gravity force.

---

9. If you are not familiar with vector algebra, the dot product here is the product of the speed of the car and the projection of the force on the direction of the velocity.

Therefore, the power required to counter the gravity force must be a function of the road incline. This function must be equal to zero when the road incline is zero[10] and will be equal to $F_g = -mg$ if the incline is equal to $\pi/2$ or when the road is going up exactly vertically.

For the problem in hand, we need to estimate the force $F_g(\alpha)$ pulling the car back on an incline α that is somewhere in between zero and $\pi/2$. To make a rough estimate, we speculate (see rule 8b) that

$$\frac{F_g(\alpha)}{F_g\left(\frac{\pi}{2}\right)} \sim \frac{\alpha}{\left(\frac{\pi}{2}\right)} = \frac{2\alpha}{\pi}, \tag{5.26}$$

where α is the road incline angle and $F_g(\alpha)$ is the projection of the force of gravity that the car engine will be working against. We already know that $F_g(\pi/2) = -mg$, which yields

$$F_g(\alpha) \sim -\frac{2\alpha}{\pi}mg. \tag{5.27}$$

Then equation (5.23) for the portion of the power spent on countering the gravity pull should be modified as follows:[11]

$$P_g = \frac{2\alpha}{\pi}mgV. \tag{5.28}$$

With these analyses in hand, we can proceed to plugging in some numbers. We use the following assumptions:

1. The car is moving at $V = 30$ m/s; this roughly corresponds to 100 km/hr or 60 mph.
2. The car's weight is assumed to be $m = 1.5 \cdot 10^3$ kg.
3. The drag coefficient depends on the shape of the body; for example, $C_D \approx 0.47$ for a sphere. For a car, it is reasonable to assume a lower value for C_D because cars are purposefully designed to have a lower air drag. We can use $C_D \approx 0.2$ as a rough guess.
4. The cross-section area of the car is assumed to be $A \sim 3$ m^2.
5. We are at a 6° road incline ($\alpha \approx 0.1$ radians).
6. The air density is $\rho \approx 1.2$ kg/m^3.

Using these numbers in equations (5.25) and (5.28) yields

---

10. This is similar to Galileo's reasoning discussed early in chapter 2.

11. The exact expression is $P_g = mgV \cos(\pi/2 - \alpha) = mgV \sin\alpha$, where $\sin\alpha$ appears because vectors $\vec{F}$ and $\vec{V}$ are not aligned in equation (5.23). Note that for small angles we can use approximation $\sin\alpha \approx \alpha$, but our rough estimate is equivalent to $\sin\alpha \sim 2\alpha/\pi$, which is within the correct order of magnitude for all angles $\alpha < \pi/2$.

$$P_a \sim 10^4 \ W,$$
$$P_g \sim 3 \cdot 10^4 \ W. \tag{5.29}$$

For the car engine power, horsepower units are more customary. As mentioned at the beginning of chapter 1, one horsepower is equal to ~745.7 W. This means that an engine must use about 13 hp to overcome the air resistance at 60 mph and about 40 hp to drive uphill on a 6° slope. The two numbers add up to the power expenditure of about 53 hp. In addition, the engine needs to run various car subsystems, further increasing the requirement for power. Note that the power of a car engine may decrease with the car's age, and designers must make an allowance for that. Altogether, an engine with $P \sim 100$ hp seems necessary if we want to be able to take on mountain roads at highway speeds.

This seems reasonable: a car with a 100 hp engine is able to climb 6° hills without substantial strain.

5.7 Summary

The complexity of many real-life problems is far beyond our abilities to solve them exactly and rigorously, so we often must resort to cutting corners. Knowing how to cut corners must be a part of a well-rounded education for any scientist or engineer (pun intended). Making order of magnitude estimates is just that: cutting corners when you have no other choice.

Luckily, different phenomena or different factors affecting real-life problems often have vastly different magnitudes. This means that some phenomena completely dominate the result, and the rest can be safely neglected. Selecting the phenomena that matter is an art, largely based on making order of magnitude estimates.

For each estimate, we consider just the most basic mechanism. We suspend looking at the subtlety and complexity of the problem. Guidelines for the art of order of magnitude estimates are listed in section 5.2. They should be applied creatively—after all, art is rarely made by using standard formulas.

Scaling and order of magnitude estimation are intimately connected:

1. On the one hand, the existence of multiple scaling behaviors is often the basis for two or more phenomena to differ by orders of magnitude. Consider an example given in section 4.2: the air drag has two mechanisms that act somewhat independently. One scales linearly with the velocity, and another one scales quadratically. For small velocities, the linear scaling beats the quadratic one, and the latter can be neglected. For large velocities, it is the other way around.

2. On the other hand, order of magnitude estimates may point to a limit on the scaling behavior. As discussed in chapter 4, scaling is rarely unrestricted. Most often the scal-

ing parameter lies in a finite range, limited by some extraneous factors. An order of magnitude estimate may be able to identify the range where a scaling remains valid.

Finally, an order of magnitude estimate may serve as a checking tool, just like the other techniques presented in earlier chapters. By its very nature, it is a crude tool: it will not flag an error if a solution is off by a factor of two or three. Yet, it is surprising how often we remain unconcerned by a result that is off by an order of magnitude or more. In these cases, a quick rough estimate can serve as a wake-up call.

Exercises

The exercises below are different from those in other chapters. A key part of making an order of magnitude estimate is planning a simple approach that uses known or easily obtainable numbers. It would defeat the purpose of this chapter to suggest a working approach to solve each problem, or even to suggest all the data you may use. Moreover, many problems can be correctly solved multiple ways. It is up to you to chart a course for a solution and to hunt for the necessary data. As a result, some or all input data may be missing in the formulation of the exercises below.

You are allowed to speculate and to use reference books and web searches to come up with the missing data needed to solve each problem. Of course, this does not mean that you can simply Google the final result—you still have to produce an estimate from some other data.

Problems with an asterisk present an extra challenge.

1. A screenwriter is tasked with writing a script for a movie about a bank heist. The initial concept calls for the burglars getting away with $5 million in cash. The screenwriter wants to be sure that the burglars can physically haul that amount. Estimate how much $5 million may weigh and what the combined volume of sacks is needed to carry this cash. Make calculations for three different cases:

    a)   All the cash is in $100 bills.

    b)   All the cash is in $20 bills.

    c)   The cash is in roughly equal numbers of $100 and $20 bills.

2. Mary has noticed that running on a treadmill is easier than running outside. She thought about this and decided that the reason for the difference is the air drag, which she has to overcome when running outside but not in the gym. To prepare for a race, Mary decided to set up the incline on the treadmill in such a way that her expended energy on the treadmill will be equal to that when running on a level trail outside. She read section 5.6 and followed the math there to compute the required incline. Assuming that Mary is of an average build and runs at the pace of 6 mph, what should the incline on the treadmill be?

3. Suppose that scientists have learned how to put a tag on individual molecules.[12] They use this technology to study the circulation of water in Earth's oceans and atmosphere. They tagged all the molecules in a cup of water and then poured this cup into the Amazon River. Now scientists periodically scoop a cup of water at various places around the globe and measure the number of tagged molecules there. Provide a rough estimate for the number of molecules in a scooped cup that would indicate that the original sample has mixed approximately uniformly with the water in the world's oceans.

---

12. The statement of this problem is not as absurd as it may seem. As the air circulates in the atmosphere, some air parcels travel to high altitudes, where they are bombarded by cosmic rays. This turns nitrogen into carbon-14 $\left(^{14}C\right)$; further circulation brings this carbon back to Earth, where it enters the global nutrient cycle. The "tagged" ^{14}C carbon is used by scientists to perform radiocarbon dating of archaeological and biological specimens. Understanding the dynamics of global air circulation is important for accurate radiocarbon dating.

4. Find information on how many lightning storms one may observe per year in a temperate climate zone of the world, or use your own experience if you live in a temperate zone.

    a) Based on this number, estimate how many lightning strikes occur per second globally.

    b) Note that simply multiplying the number of strikes that one may observe by the total number of people on the planet is not a valid approach. Why?

5. Every year, a number of people are injured by lightning strikes. Assume that most people are within their homes during a thunderstorm and that, if lightning strikes a house, its inhabitants are injured. Using the results from the previous problem, estimate the number of people who are injured by lightning annually.

6. How many jelly beans can fit in a 2 liter jar?

7. US counties draw their revenue primarily from real estate taxes. Select a particular county and estimate its budget. Assume that the annual tax rate is 1 percent of the cost of each house.

8. The Stefan–Boltzmann law predicts the power of total electromagnetic radiation (including light) emitted by a heated body from a unit area:

$$P = \sigma T^4, \tag{5.30}$$

where $\sigma \approx 5.67 \cdot 10^{-8}$ W/(m^{2}K^4). Here the power is measured in watts (W) and temperature T is in the Kelvin scale.

    a) Use the size of the Sun and its surface temperature of 6,000 K to estimate the total energy emitted per second.

    b) Use the scaling that was discussed in section 4.5 to determine the power per unit area away from the Sun at Earth's orbit.

    c) Find reference information on the power that can be produced by solar panels on the ground per unit area. How does this number compare with the total power of sunlight per unit area?

9. According to a popular legend, the inventor of the game of chess came to the local king and showed him the game. The king loved the game and asked the inventor to name a reward. The inventor said that he wanted one grain of rice for the first square on the board, two grains of rice for the second square, four grains for the third square, and so on. The king, apparently not being very good at math, quickly agreed.

    a) Knowing that a chessboard has 64 squares, estimate the total weight of rice that was to be received by the inventor.

    b) How does it compare with today's global annual rice production?

10. A spherical cap is the part of a sphere that lies above a plane that crosses this sphere (see section A.30 and figure 5.2).

    a) Produce order of magnitude estimates for the volume and the surface area of the spherical cap as expressed through h and R. (Hint: For the surface area, approximate the shape of the

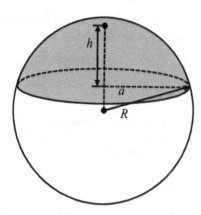

Figure 5.2
A spherical cap: order of magnitude estimates

        cap as a disk with radius a. For the volume, approximate the shape of the cap as a cylinder with radius a and height h. Then use the Pythagorean theorem to solve for a through h and R.)

b) Try to improve your estimates by approximating the shape of the cap as a cone instead of a cylinder. The volume and the area of a cone (net of the area of the base) are given by
$V_{cone} = \pi a^2 h/3; \quad S_{cone} = \pi \sqrt{h^2 + a^2}$.

c) Compare the two sets of estimates with the exact formulas in section A.30.

11. In the seventeenth century, Peter Minuit orchestrated the notorious purchase of Manhattan from a native American tribe.

a) Estimate the total value of real estate in Manhattan in today's prices.

b) Using this value, estimate the total current cost of land on the island.

c) Assume that the purchase occurred in 1626 and the land was purchased for the equivalent of $1,000 in today's money. Estimate the rate of return for the original purchase price to grow to the value of the Manhattan land today. This assumes that a value grows exponentially in time as $I = I_0 e^{rt}$, where I_0 is the initial amount and r is the rate of return.

d) How would the estimate for the rate of return change if we assume that Peter Minuit paid $24 for the purchase?

e) How does it compare with the 7–8 percent growth that many people expect to see in their retirement accounts?

12. A torus is a donut-shaped body (see figure 5.3). Use an order of magnitude estimate to flag the incorrect formulas for the volume of a torus below. (Hint: Estimate the volume of a torus as

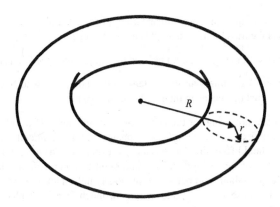

Figure 5.3
A torus: order of magnitude estimates

the volume of a cylinder you would get if you cut the torus at one place and unrolled it to a cylinder.)

a) $V = 2Rr^2$

b) $V = 2\pi^2 Rr^2$

c) $V = 12\pi^2 Rr^2$

13. A powerful explosion creates a supersonic[13] spherical shock wave. As the wave propagates outward, it slows down, and at some point its speed falls below the speed of sound. Use equation (4.38) in section 4.8 to estimate the maximum radius of the supersonic shock wave from an explosion caused by 10 kg of TNT that releases the energy $E \approx 4 \cdot 10^7$ kg $\cdot$ m$^2 \cdot$ s^{-2}.

14. The velocity of gravity waves in deep water (in the case of $\lambda \ll h$) and in shallow water (in the case of $\lambda \gg h$) is given by (see problem 15 in chapter 4):

$$V_{\text{deep}} = \sqrt{\frac{g\lambda}{2\pi}},$$

$$V_{\text{shallow}} = \sqrt{gh},$$

(5.31)

where λ is the wavelength (the distance between two consecutive wave crests), h is the depth of the water basin, and $g \approx 9.8$ m/s^2 is the gravity acceleration. Tsunamis are large gravity waves in the ocean that are often caused by underwater earthquakes. A tsunami wavelength can be on

---

13. A shock wave is supersonic if it travels at a speed that exceeds the speed of sound.

the order of the size of the region on the ocean floor affected by the earthquake, or about 10^2 km.

An engineer is tasked to come up with a preliminary design of a tsunami warning system. The key factor in the design is the travel time of tsunamis across the ocean.

a) Which of the equations in system (5.31) should be used to estimate the speed of a tsunami?

b) How much in advance can a satellite warning system notify the people who live on the Pacific Rim for a tsunami that originates in the middle of the ocean?

15. Zoo employees Jim and Jack are preparing for the arrival of their first adult African bush elephant to the zoo. They have no experience caring for elephants and are frantically trying to figure out what and how they should feed him. Jim has a lot of experience in caring for rabbits and knows an excellent supplier of rabbit food. After some brainstorming, Jim and Jack decide that the elephant might enjoy rabbit food but would obviously need a lot of it. They apply Klieber's law (see equation (4.2) in section 4.1) to estimate how much food an elephant needs. What estimate may they come up with?

16. The radius r of Archimedes's spiral increases linearly with the angle θ of the turn (figure 5.4). Mathematically, this is expressed in polar coordinates as

$$r = a\theta. \tag{5.32}$$

Emma found a formula on the web for the arc length of the spiral:

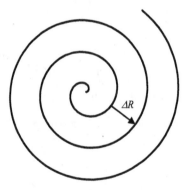

Figure 5.4
Archimedes's spiral: order of magnitude estimates

$$L = \frac{a}{2}\left(\theta\sqrt{1+\theta^2} + \ln\left(\theta + \sqrt{1+\theta^2}\right)\right). \tag{5.33}$$

She knows that the spiral is fully defined by two parameters: the total turn angle (dimensionless) and the distance between the adjacent loops (measured in meters). By applying dimensional analysis to equation (5.33), Emma concluded that a may be denoting the distance between the adjacent loops. To check this conjecture, she decided to compare equation (5.33) to a rough estimate for the arc length. Reproduce Emma's estimate for the arc length of the Archimedes's spiral and analyze the result:

a) Derive a rough estimate for the arc length of the spiral as expressed through the total turn angle θ and the distance between adjacent loops ΔR. (Hint: First estimate the number of loops, and then use rule 4 in section 5.2 to estimate the length of each loop.)

b) Compare this estimate with equation (5.33), assuming that parameter a in that equation denotes the distance between adjacent loops: $a = \Delta R$. Is your rough estimate consistent with that formula?

c) Does your rough estimate include a logarithmic term that is a part of equation (5.33)? If not, is it important?

d) Refer to a more complete formulation in section 6.6. What is the correct relationship between a and ΔR?

17. Estimate the length of the curve defined by $y = \sin x$ for $0 \le x \le 2\pi$. Compare your result with $L \approx 7.6404$.

18*. Consider the curve that is a plot of function $y = bx^2$.

a) Estimate the length of this curve for $0 \le x \le c$.

b) Compare your estimate with the exact result:

$$L = \frac{c}{2}\sqrt{4b^2c^2 + 1} + \frac{1}{4b}\ln\left(2bc + \sqrt{4b^2c^2 + 1}\right). \tag{5.34}$$

c) Does your rough estimate include a logarithmic term? If not, is it important?

d) Plot your estimate and the exact formula for $b = 2$ as a function of c.

19. Equation (2.94) in section 2.12 estimates the maximum possible radius of coverage by a low-orbit satellite that transmits a signal to customers on Earth's surface. Assume you are planning a satellite constellation that must cover the entire globe and that these satellites will fly at an altitude of 400 km. How many satellites will you need?

20. The Greenland ice sheet covers about $1.7 \cdot 10^6$ km² and is 3 km thick at the thickest point. Give a rough estimate for the rise of the sea level if this ice sheet completely melts as a result of climate change.

21. Estimate $n! = 1 \times 2 \times \cdots \times n$ as a function of n. Check the quality of this estimate for $n = 5$ and $n = 10$. (Hint: Since $n!$ is a product of n factors, use the nth power of the average value for these factors.) Compare the functional form of your estimate to the Stirling formula, which approximates the factorial as

$$n! \approx \sqrt{2\pi n}\left(\frac{n}{e}\right)^n.$$ (5.35)

Note that a rough estimate may miss the Stirling formula value or the exact value by more than one order of magnitude for a large n but still captures the rough behavior of $n!$. The reason for the crudeness of this estimate is the extremely steep dependence of $n!$ on n.

22. To put a satellite into orbit, we need to design a booster. The Tsiolkovsky rocket equation (see section 4.2) links the velocity of the exhaust V_e, the total velocity gained by the rocket V, the total starting mass of the rocket including propellant m_0, and the final mass of the rocket when the propellant has been used m_f:

$$V = V_e \ln \frac{m_0}{m_f}.$$ (5.36)

The velocity of the exhaust leaving the rocket depends on the type of fuel used and on engineering constraints; for this problem, use $V_e = 2,000$ m/s. With these data in hand, start planning a satellite mission:

a)   The radius of a low-Earth orbit is only slightly larger than Earth's radius. Use the results from problem 4 in chapter 1 to estimate the satellite velocity on the orbit.

b)   Use the Tsiolkovsky equation (5.36) to estimate the total mass of a rocket that is needed to put a 10-ton satellite on a low-Earth orbit.

c)   Compare your estimates with masses of real launch vehicles that have been used for launching satellites on low-Earth orbits, such as *Zenit-2* or *Falcon 9 v1.0*.

23. In statistics and probability theory, the so-called logistic distribution is given by

$$\tilde{P}(x) = \frac{c}{\left(e^{\frac{x}{2}} + e^{-\frac{x}{2}}\right)^2},$$ (5.37)

where constant c is chosen so that the total area under the curve defined by equation (5.37) is equal to 1. Give an estimate for c.

24. If an object is freely falling in air, its velocity increases until the force of gravity becomes balanced by the air drag force. This steady fall velocity is called the terminal velocity (see also problem 25 in chapter 4). Depending on the values of the parameters, the terminal velocity for a sphere may be given by one of the following equations:

$$mg = \frac{1}{2}C_D\rho A V_t^2,$$
$$mg = 6\pi R\eta V_t,$$ (5.38)

where g is the gravity acceleration, m is the mass of the object, V_t is the terminal velocity with respect to the air, R is the radius, A is the cross-section area, C_D is the drag coefficient ($C_D \approx 0.47$ for a sphere), and η is called the dynamic viscosity, a property of the medium. For air, assume $\eta = 1.8 \cdot 10^{-5}$kg/(m · s). Consider a hailstone that grows slowly in a cloud. At any time, the velocity of its fall through the air is approximately equal to the terminal velocity. As the size and the speed of the hailstone increase, its fall transitions between the two regimes for the air drag. Estimate the size of the hailstone at which this transition occurs.

25*. Linear regression is an algorithm that is commonly used to interpret numerical data (see section A.33). Assume that we have N pairs of measurements x_i and y_i for variables x and y. The underlying linear relationship between these two variables is corrupted by measurement noise R:

$$y = ax + b + R. \tag{5.39}$$

The linear regression algorithm estimates the best fit for model parameters a and b from available data:

$$a = \frac{N \sum_{i=1}^{N} x_i y_i - \sum_{i=1}^{N} x_i \cdot \sum_{i=1}^{N} y_i}{N \sum_{i=1}^{N} x_i^2 - \left(\sum_{i=1}^{N} x_i\right)^2},$$

$$b = \frac{\sum_{i=1}^{N} y_i \cdot \sum_{i=1}^{N} x_i^2 - \sum_{i=1}^{N} x_i \cdot \sum_{i=1}^{N} x_i y_i}{N \sum_{i=1}^{N} x_i^2 - \left(\sum_{i=1}^{N} x_i\right)^2}, \tag{5.40}$$

where x_i, y_i are the available data for variables x and y.

For this problem, assume that the data points form two approximately equal clusters. The first cluster is centered at $(x_1 = 1, y_1 = 2)$, and the second cluster is centered at $(x_2 = 6, y_2 = 4)$. In each cluster, points are located within a distance less than 2 units from the cluster center. Compute an estimate for parameters a and b of a linear fit to this data set. (Hint: Use rule 7 in section 5.2 to estimate various sums in equations (5.40).)

6 Successive Approximations

In the beginning of the nineteenth century, astronomers published data on the orbit of the last known planet at the time, Uranus. The observations showed irregularities in the orbit that could not be explained by the gravity from the Sun and from the other planets. Urbain Le Verrier and John Couch Adams independently embarked on investigations to check if these irregularities could be caused by the gravity from another, yet unknown planet.

The equations of motion for a gravitational system that includes three or more bodies are very complex, and there was no chance of getting a neat, closed-form solution that accounts for the Sun and multiple planets. Fortunately, researchers had a trump card to use: the Sun is by far the most massive body in the solar system, and it enacts by far the strongest force on each planet. The interplanet attraction is relatively weak. If it is neglected completely, each orbit is elliptical and well known. Starting from that foothold and from some rough initial guess for the orbit of the new, hypothetical planet,[1] one can come up with approximate equations, which yield a correction to that initial guess. The corrected initial guess would then become close to the exact solution for the orbit of the new planet, assuming that this planet exists. This is exactly what Le Verrier and Adams did.

Mathematical predictions for the orbit of a new planet allowed astronomers to target their telescopes to the correct region of the sky. In a triumph for Newtonian mechanics and math, a new planet was indeed discovered there. It was called Neptune.

That scientific breakthrough shows the power of using a small parameter[2] as leverage for solving tough problems. Very few real-life problems permit an exact, closed-form solution. There may be multiple phenomena of different magnitude that affect the final result, complicating the problem and placing an exact solution beyond reach. In chapter 5 we learned how to roughly estimate the magnitude of different quantities; often this helps select the

---

1. For an initial rough guess on the orbit of the new planet, Adams used the Titius–Bode law, which states that the orbit radius of each next planet is approximately double that of the previous planet. By doing that he applied scaling!

2. In the Neptune discovery case, the masses of planets were treated as small compared to the mass of the Sun.

phenomena that can be neglected entirely. Unfortunately, the remaining phenomena may still be too complex for getting a closed-form solution. In these cases, the next best thing to *neglecting* a weak effect entirely is *using* its small magnitude to get an approximate result.

In this chapter we will learn a basic version of a powerful problem-solving method based on this tactic. The point of this chapter is that, if you cannot neglect a small effect, use its size to your benefit.

We have already formalized the notion of "small" and "large" quantities (see chapter 1 and section 5.1). As a reminder, we refer to parameter μ as small if it is dimensionless and if $|\mu| \ll 1$, where notation $\ll$ was defined by equations (5.3) and (5.4). Often it is sufficient for μ to be one order of magnitude less than 1, and if it is two or more orders of magnitude less than 1, the "smallness" criterion becomes stronger.

The most common way to exploit a small parameter is to apply the *method of successive approximations* (MSA).[3] The idea is to build a solution step by step, often starting from the case when the small parameter is set to zero and then extending this solution to nonzero values.[4]

Key Point

In practice, very few problems yield a closed-form solution. Frequently a way to make inroads into a tough problem is to exploit small parameters there. Small parameters naturally appear in applications because some phenomena have a smaller (often a much smaller) magnitude than others.

Here you may detect the whiff of using the value of zero as a limiting case for the small parameter in question (see chapter 2 for the discussion of limiting cases). Indeed, it is common to follow a three-step procedure for solving a tough problem:

1. Identify a dimensionless small parameter.
2. Solve the problem in the limiting case when the value of this parameter is set to zero.
3. Use the MSA to extend the solution to nonzero (but still small) values of the small parameter.

Like any approach for solving mathematical problems, this one is not universal. A small parameter may not exist, a limiting case may be too difficult to crack, or the MSA may fail. Yet, this approach has been instrumental for countless problems that resist any other solution.

---

3. Depending on the particular flavor, the MSA may be referred to as the perturbation theory, the asymptotic method, or the small parameter expansion.

4. In its full-fledged form this method uses calculus, but some applications require only basic knowledge of algebra. In this book we limit the treatment to the latter.

In the following sections we will learn how to apply successive approximations using a few sample problems.

6.1 Achilles and the Tortoise

We start from a problem that seems simple today but created a controversy about 2,500 years ago. Zeno, a Greek philosopher, argued that Achilles, who was known for his fast running, could never win a foot race against a tortoise. Zeno's argument (modernized and reworded here) was as follows:

1. In the beginning, the tortoise has a head start of D_0 meters. Both Achilles and the tortoise start running at the same time.
2. While Achilles runs D_0 meters, the tortoise runs D_1 meters. The racers are now separated by D_1 meters. Note that $D_1 < D_0$ because the tortoise is slower than Achilles.
3. While Achilles runs this new segment of D_1 meters, the tortoise runs another distance D_2. Achilles is still behind the tortoise.
4. This sequence can be extended indefinitely. According to Zeno, it would take an infinite number of stretches for Achilles to reach the tortoise, which would mean that he could never do it.

This argument is called Zeno's paradox. Below we present two solutions to this problem, one of which will be used to introduce the MSA.

6.1.1 First Solution

Here we depart from Zeno's argument and use a modern algebraic approach. We assume that Achilles runs with speed V_A and the tortoise runs with speed V_T. We also assume that $V_T < V_A$. The positions of Achilles and of the tortoise as functions of time t are described by the following equations:

$$x_A = V_A t,$$
$$x_T = V_T t + D_0. \tag{6.1}$$

Achilles reaches the tortoise when $x_A = x_T$ (see figure 6.1). Using this condition in equations (6.1) and solving for time yields

$$T = \frac{D_0}{V_A - V_T}, \tag{6.2}$$

where T denotes the time that is required for Achilles to reach the tortoise. This completes the algebraic solution.

6.1.2 Second Solution

In this solution, we faithfully follow Zeno's argument to arrive at the same result:

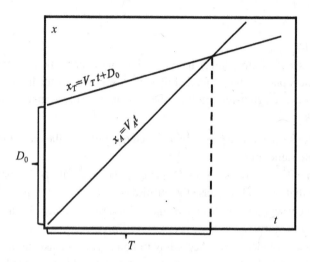

Figure 6.1
Achilles and the tortoise: an algebraic solution

1. In the first segment of the race, Achilles runs distance D_0. At his speed, it takes him

$$T_0 = \frac{D_0}{V_A} \tag{6.3}$$

seconds to do that.

2. The tortoise uses this time to move farther. At her speed, she runs the following distance in T_0 seconds:

$$D_1 = V_T T_0 = D_0 \frac{V_T}{V_A}. \tag{6.4}$$

3. In the next segment of the race, Achilles runs D_1 meters. It takes him

$$T_1 = \frac{D_1}{V_A} = \frac{D_0}{V_A} \cdot \frac{V_T}{V_A} \tag{6.5}$$

seconds.

4. This sequence can be extended indefinitely. The duration of every next segment is obtained from the duration of the previous segment by multiplying it by V_T/V_A.

The total time is therefore given by the sum of the durations of all segments:

$$T = \frac{D_0}{V_A}\left(1 + \frac{V_T}{V_A} + \left(\frac{V_T}{V_A}\right)^2 + \left(\frac{V_T}{V_A}\right)^3 + \cdots\right). \tag{6.6}$$

In the last equation, we recognize a geometric series. Using the formula for the sum of a geometric series,[5] we get

$$T = \frac{D_0}{V_A} \cdot \frac{1}{1 - \frac{V_T}{V_A}} = \frac{D_0}{V_A - V_T}. \tag{6.7}$$

Predictably, this is the same result as in equation (6.2).

6.1.3   The Method of Successive Approximations

Zeno's way of treating this problem is peculiar. It is a sequence of conceptually identical steps, each getting ever closer to the exact solution. At each step, we must deal with two phenomena: Achilles's and the tortoise's running, but instead of considering them jointly, we alternate between them. This decoupling of the two phenomena makes the solution at each step inaccurate, but the bulk of the error is corrected at the next step.

Let us consider how this technique works graphically. Figure 6.2 shows the sequence of the steps we took. Let us walk through these steps again:

1. In the first segment of the race, Achilles runs distance D_0, as if the tortoise were standing still. It takes T_0 seconds for him to do that (see equation (6.3)). In figure 6.2, this step is shown by the arrow from point K to point L, which spans the time interval $0 \le t < T_0$.

2. Next, we consider the movement of the tortoise. During time $0 \le t < T_0$, the tortoise moves from her starting point at $x = D_0$, covering distance D_1 (see equation (6.4)). She ends at position $x = D_0 + D_1$. This is shown by the arrow from point L to point M.

3. In the next segment, Achilles runs distance D_1 in T_1 seconds (see equation (6.5)). This is shown by the arrow from point M to point N. The total time that has elapsed since the start of the race is given by $T_0 + T_1$. Achilles ends up at the position and at the time moment that correspond to point N in the figure.

4. These steps can be repeated indefinitely. The sequence of arrows quickly approaches the point of the intersection of the two lines that depict the motion of Achilles and of the tortoise. The intersection point is when and where Achilles reaches the tortoise.

The idea of getting step by step ever closer to the exact solution is the crux of the MSA. We say that successive approximations *converge* to the exact solution.

---

5. For $|x| < 1$, the sum of a geometric series is given by $1 + x + x^2 + x^3 + \cdots = 1/(1 - x)$.

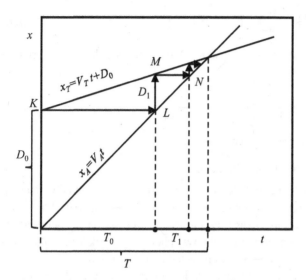

Figure 6.2
Achilles and the tortoise: an iterative solution

In practice, the result of this method is usually truncated after getting a few terms or even just the first term. Time T_0 would be called the zero-order approximation, time $T_0 + T_1$ the first-order approximation, and so on. At any given order, the result is approximate. Note that the MSA is different from many computer algorithms that produce a numerical approximation to a solution: it yields a formula rather than a number. This formula can be analyzed, studied, and played with, for example, using the methods we explored in earlier chapters.[6]

Before we proceed further, we make a few observations:

1. This method requires having a small dimensionless parameter, $|V_T/V_A| < 1$. The smaller the parameter, the faster the convergence to the exact solution. Without a small parameter, this method does not work.

2. Often, but not always, successive approximations exhibit power law scaling with respect to the small parameter; that is, each next approximation contains a progressively

---

6. This is not unlike any solution expressed through trigonometric or other transcendental functions. Indeed, a numerical value of such a solution is usually obtained from an iterative algorithm and is approximate. The advantage of expressing a solution through a trigonometric function, rather than just giving a number, is that these functions have many well-studied analytical properties, which helps with manipulating the result and with the analysis.

higher power of the small parameter. The problem of Achilles and the tortoise is a pure example of that: it produces a geometric series.

3. If a problem deals with two phenomena, the MSA often treats them separately and in turn. In the example above, we consider Achilles reaching the position of the tortoise as if the tortoise were standing still, and only then do we consider the effect of the tortoise moving.

4. The MSA does not always converge. For the Achilles and the tortoise problem, the convergence criterion is that the tortoise must run slower than Achilles. However, suppose that the race happens after Achilles was injured in his heel by Paris[7] and was barely able to walk. If the tortoise runs faster than Achilles, the successive approximations do not work at all. On the other hand, the algebraic solution (equation (6.2)) will simply produce a negative sign for T, indicating that the tortoise may have passed Achilles before the race started.

5. In the Achilles and the tortoise example, the zero-order approximation is a limiting case for $V_T \to 0$. Commonly, we start from a limiting case for the zero value of the small parameter and then build a solution from it.

6. If you are familiar with calculus, you may recognize that the geometric series in equation (6.6) is a Taylor series for the exact solution (6.2). This is often the case, but the MSA is not always equivalent or even related to the Taylor expansion.

Now we can move from this rudimentary example to a more general explanation for this method.

6.2 How MSA Works

How can we generalize the iterative solution of the Achilles and the tortoise problem to more complex and realistic problems? Here is one common scenario that illustrates this method but does not exhaust all possibilities for applying it. The following qualitative argument is not rigorous and occasionally may not work in practice. However, it works amazingly well for a vast variety of problems.

Often a researcher starts from a known problem and extends it by accounting for additional phenomena. Suppose a problem with a known solution is given by the following equation:

$$F(x) = a. \tag{6.8}$$

Its solution is

$$x = F^{-1}(a), \tag{6.9}$$

---

7. In this story, Paris is a person, not the city.

where $F^{-1}(a)$ is the inverse function of $F(x)$. A new problem modifies the original equation:

$$F(x) = a + b \cdot G(x), \tag{6.10}$$

where function $G(x)$ models the new phenomena. In many cases, the new term $G(x)$ can be complex and would prevent any attempts to get a closed-form solution for the new equation. Even so, the original problem is a limiting case of the modified one for $b \to 0$. This gives us an opportunity to gain on the modified problem for small values of b.

We "solve" the modified problem following the same procedure as before to get

$$x = F^{-1}(a + b \cdot G(x)). \tag{6.11}$$

This is not a solution in the conventional sense because the unknown is still present in the right-hand side. However, if parameter b is "small," the last equation may offer a convenient and relatively simple way to approach the solution.

We start from some initial guess x_0 for x, which is called the *zero-order approximation*. Quite often it is the solution in the limiting case $b = 0$. Starting from this value, we build a sequence of approximations for x:

$$x_n = F^{-1}(a + b \cdot G(x_{n-1})). \tag{6.12}$$

Here x_n is *defined* by equation (6.12), so equation (6.12) is exact, in the sense that the left-hand side exactly equals the right-hand side. The next approximation x_1 is called the *first-order approximation* and so on.

At the same time, equation (6.12) is approximate in the sense that it approximates the original equation (6.11):

$$\begin{aligned} x_n &\approx x, \\ F^{-1}(a + b \cdot G(x_{n-1})) &\approx F^{-1}(a + b \cdot G(x)). \end{aligned} \tag{6.13}$$

Since both equations (6.13) are approximate, there is a (generally) nonzero difference between the left-hand side and the right-hand side in each. To quantify this difference, we subtract equation (6.12) from equation (6.11):

$$x - x_n = F^{-1}(a + b \cdot G(x)) - F^{-1}(a + b \cdot G(x_{n-1})). \tag{6.14}$$

Now let us look at equations (6.12) through (6.14). We note the following:

1. The left-hand side of equation (6.14) is the approximation error for the left-hand side in equation (6.12). The term "approximation error" here is understood in the sense that x_n approximates x, so that $x - x_n$ is the difference between the exact and the approximate values.

2. The right-hand side of equation (6.14) is the approximation error for the right-hand side in equation (6.12). The term "approximation error" here is used in the sense that $F^{-1}(a + b \cdot G(x_{n-1}))$ approximates $F^{-1}(a + b \cdot G(x))$.

3. Per equation (6.14), the approximation errors for the left-hand and the right-hand sides of equation (6.12) are exactly equal.

Yet, in the right-hand side of equation (6.12) the source of this error (which is ultimately caused by $x_{n-1} \neq x$) is multiplied by a small parameter b. This reduces the effect of any approximation error in x_{n-1}: even if the error in x_{n-1} is relatively large, its effect on the right-hand side of equation (6.12) will be mitigated. In the left-hand side there is no small parameter, and the effect of the approximation error in x_n is not reduced. To compensate for the error reduction in the right-hand side, and to have the errors in both sides of equation (6.12) equal, we should expect x_n to be more accurate than x_{n-1}. The last statement is key: it is the mechanism that makes each next approximation closer to the exact solution.

Most important, we can repeat this procedure indefinitely and get a chain of ever more accurate approximations! For the simple problem of Achilles and the tortoise, this technique seems quite cumbersome and unnecessary, but for many challenging problems it becomes the only way to obtain a solution.

The explanation above is not rigorous and would fail the high standards of a true mathematical derivation. This is a compromise born of necessity: a truly rigorous, fail-proof solution method for many real-life problems simply does not exist. In these cases the MSA often delivers a practical result. After all, a mathematician solves a problem that she can solve, in a way that must be used there; and an engineer solves the problem that must be solved, in a way she can use there.

In this section, we considered a case where the term with the small parameter is additive. This is not a strict requirement for applying the MSA. This method has many different versions and flavors, and we explore one example of a case with a nonadditive small term in section 6.7.

6.3   When It Works and When It Doesn't

Section 6.1 discussed that the MSA works for the Achilles and the tortoise problem only if $|V_T/V_A| < 1$. In this section, we address a more general question: what is the criterion for the MSA to converge for the type of problems considered in section 6.2?

A proof of this criterion is beyond the scope of this book, and we examine only the result. We also note that a rigorous application of this criterion is not always feasible, and researchers often rely on a hunch: they call it a success when they see that the first few iterations get ever closer to some value. Of course, a mathematical purist would correctly observe that an initial "success" does not guarantee the convergence of subsequent iterations.

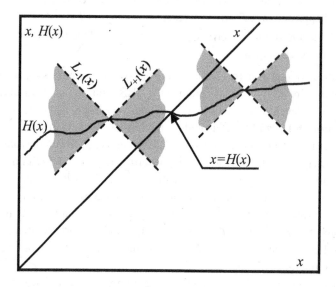

Figure 6.3
Convergence criterion for the MSA

We start from equation (6.11). If we denote $H(x) = F^{-1}(a + b \cdot G(x))$, that equation takes the form

$$x = H(x). \qquad (6.15)$$

Now the application of the MSA is reduced to iterations on this equation (see section 6.2):

$$x_n = H(x_{n-1}). \qquad (6.16)$$

Figure 6.3 notionally shows the plots of x and $H(x)$ as functions of x. The arrow points to the solution of equation (6.15). Assume now that we draw lines with slopes -1 and $+1$ at *every point* of the plot of $H(x)$. These lines are given by equations $L_{+1}(x) = H(x_p) + x - x_p$ and $L_{-1}(x) = H(x_p) - x + x_p$ for any point $(x_p, H(x_p))$ on the plot of function $H(x)$. Each line pair will define a region between these lines.

Only two such pairs of lines are shown in figure 6.3, and the corresponding regions are shaded there. If the plot of $H(x)$ lies entirely within all these shaded regions, the MSA will converge.[8] The opposite is not true: even if the plot of $H(x)$ sticks out of some of the shaded regions, the MSA still *may* converge, but there is no guarantee of that.

If you are familiar with calculus, another criterion may come in handy. If function $H(x)$ is differentiable and if its derivative satisfies inequality $\left|\frac{dH(x)}{dx}\right| < 1$ everywhere, then the MSA will converge.

Key Point

Watch for situations when the MSA does not converge. It's not always possible to rigorously apply convergence criteria, and the impression of convergence from the first couple of iterations may be deceptive.

The discussion on convergence criteria would be incomplete without mentioning computer implementations of the MSA. Unlike Urbain Le Verrier and John Couch Adams, researchers of the modern era routinely implement MSA iterations in computer software. The finite precision of machine operations introduces additional possibilities for the MSA to diverge. The art of designing robust numerical algorithms is a separate discipline and is beyond the scope of this book.

Below we go through several examples that show how the MSA concept works in practice. This will include some problems that are difficult or impossible to solve using any other technique.

6.4   The Product of Two Linear Expressions

We start from a problem that has a closed-form solution (see section A.7):

$$(x - a)(x - b) = d. \tag{6.17}$$

The two solutions are

$$x_{\pm} = \frac{(a + b) \pm \sqrt{(a - b)^2 + 4d}}{2}. \tag{6.18}$$

Here we approach this problem using the MSA, assuming that d is "small." The exact criterion for the smallness of d will be evident from the final result.

We isolate an easily solvable expression in the left-hand side and lump everything else with the small parameter. Specifically, we transform equation (6.17) into[9]

---

8. The rigorous mathematical criterion states that function $H(x)$ is Lipschitz continuous, with the Lipschitz constant $K < 1$.

9. The choice of isolating $x - a$ and then solving for x is arbitrary; doing this for $x - b$ would be similar because of the $a \leftrightarrow b$ symmetry in this problem.

$$x = a + \frac{d}{x - b}.$$ (6.19)

We use a limiting case $d \to 0$ as a starting point.[10] This gives us a zero-order approximation:

$$x_0 = a.$$ (6.20)

For the first-order approximation, we plug the value of x_0 in the right-hand side of equation (6.19). The intuitive reasoning here is that the ratio in the right-hand side is already small compared to a because d is small; therefore, using an inaccurate value for x there would not have a big impact on the total:

$$x_1 = a + \frac{d}{x_0 - b} = a + \frac{d}{a - b}.$$ (6.21)

This process can be repeated indefinitely; the second-order approximation is

$$
\begin{aligned}
x_2 &= a + \frac{d}{x_1 - b} \\
&= a + \frac{d}{a + \frac{d}{a-b} - b} \\
&= a + \frac{d(a - b)}{(a - b)^2 + d}.
\end{aligned}
$$ (6.22)

Let us compare these approximations with the exact solution (6.18), which is fortunately available for this problem. For $a = 5, d = 0.5$, table 6.1 shows exact solutions x_+, x_-, and the two first approximations x_1, x_2 for several values of parameter b.

Table 6.1
Successive approximations for the product of two linear expressions

b	x_+	x_-	x_1	x_2
0	5.09807	−0.09807	5.1	5.098039216
1	5.12132	0.87867	5.125	5.121212
4	5.36602	3.63397	5.5	5.33333
6	6.366025	4.63397	4.5	4.66667
9	9.12132	4.878679	4.875	4.878788
10	10.098076	4.90192	4.9	4.90196

10. We have already explored this limiting case in section 2.1.

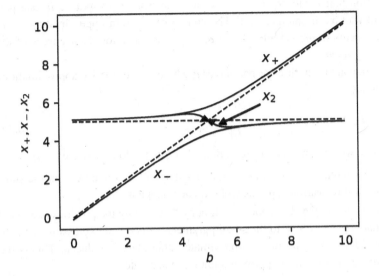

Figure 6.4
The product of two linear expressions: successive approximations

We can see three patterns:

1. The results for successive approximations cluster at only one of the two roots. Specifically, they approximate the root, which is close to the value of a. This is understandable: we started from equation (6.19) and proceeded under the assumption that the second term in the right-hand side is small, which means that the right-hand side is close to a. Of course, we can get approximations for the second root in a completely analogous way by solving for $x - b$ first.

2. The second-order approximation is more accurate than the first-order one.

3. The approximation accuracy is worse for $b = 4$ and $b = 6$ and is better for $b = 0, 1, 9$, and 10. The pattern here is that the approximation accuracy worsens when $b \approx a$ (as a reminder, $a = 5$ for the results in the table). If we look at equation (6.19), we will see that for $b \approx a$, the absolute value of the ratio in the right-hand side can no longer be assumed to be small,[11] even if d has a small value. We conclude that the small parameter assumption is eroded for $b \approx a$, affecting the accuracy of the result.

The last point is further illustrated in figure 6.4. We plot two exact solutions x_+ and x_- and the second-order approximation x_2 as functions of b. The second-order approximation

---

11. Moreover, for $a = b$, the first-order approximation does not exist because of a zero in the denominator. Since the second-order approximation is obtained from the first-order one, it is not valid for $a = b$ either.

clings to the solution that is closer to $x = a$, which corresponds to root x_+ in one part of the plot and to root x_- in another part. The switch of x_2 from approximating x_+ to approximating x_- occurs in the vicinity of $b = a$. Near this transition the accuracy of the approximation is degraded.

Now we can formulate a criterion for parameter d where the successive approximations yield a good accuracy:

$$\left| \frac{d}{a-b} \right| \ll |a|. \tag{6.23}$$

From this problem we can draw two observations that often help in practice:

1. The criterion for having a valid small parameter may not be obvious from the start. Sometimes we just try various options and see what happens.

2. At each step, we obtain the next approximation by substituting the previous one in the expression that is proportional to the small parameter. Since that expression is small, any error caused by plugging in an approximate value there is reduced. This is the mechanism that makes the next approximation more accurate.

This example allows a closed-form solution, as does the Achilles and the tortoise problem, and computing successive approximations is not necessary. Yet, it is easy to ratchet up the difficulty of a problem so that a closed-form solution becomes impractical and the successive approximation method shines. Consider the following equation:

$$(x - a)(x - b)(x - c) = d. \tag{6.24}$$

The expansion of the left-hand side will produce a cubic equation, which does have a closed-form solution. Unfortunately, the cubic formula (see section A.29) is very cumbersome and is rarely used. Yet, for small values of $|d|$ we can easily get a good approximation for each root! This is the subject of one of the exercises for this chapter.

Before we move to the next example, we note that equations like (6.17) and (6.24) surface in many different areas. For example, suppose that there are two types of waves propagating in a medium. If x is the frequency of these waves, and parameters a and b are functions of the wavelength, equation (6.17) would describe wave interaction. When a and b are substantially distinct, the waves propagate nearly independently, but they enter a resonance when $a \approx b$. Similarly, equation (6.24) may describe three different types of waves in a medium and a coupling between the different waves when any two of the parameters a, b, and c are close.

6.5   The Quadratic Equation

The MSA is usually enabled by a small parameter, but in some cases the role of the small parameter can be played by the very unknown we are solving for. In this section we investigate a simple problem for which we have already checked limiting cases. One of these limiting cases will be the starting point in the application of the MSA here. We go back to the quadratic equation

$$ax^2 + bx + c = 0. \tag{6.25}$$

In section 2.4 we investigated a limiting case $c = 0$. We concluded that one of the roots is equal to zero at the limiting case. Here we leverage this limiting case for getting a solution for small, nonzero values of c. It is a good bet that for small values of c, one root of equation (6.25) is also going to be small. As we already know from the scaling hierarchy (see section 4.2), the linear term dominates the quadratic term for small values of $|x|$. Therefore, we can treat the quadratic term here as a small perturbation. We solve equation (6.25) for x as if the quadratic term were negligible:

$$x = -\frac{c + ax^2}{b}. \tag{6.26}$$

The zero-order approximation corresponds to the already established limiting case $c = 0, x_0 = 0$. All subsequent approximations are obtained by plugging the previous approximation in the right-hand side of the last equation:

$$
\begin{aligned}
x_1 &= -\frac{c + ax_0^2}{b} = -\frac{c}{b}, \\
x_2 &= -\frac{c + ax_1^2}{b} = -\frac{c + a\frac{c^2}{b^2}}{b} = -\frac{c}{b} - \frac{ac^2}{b^3}, \\
x_3 &= -\frac{c + ax_2^2}{b} = -\frac{c + a\left(-\frac{c}{b} - \frac{ac^2}{b^3}\right)^2}{b} = -\frac{c}{b} - \frac{ac^2}{b^3} - \frac{2a^2c^3}{b^5} - \frac{a^3c^4}{b^7},
\end{aligned} \tag{6.27}
$$

and so on.[12]

Figure 6.5 shows approximations for orders 1 through 6 as a function of c for $a = 1; b = -2$. For smaller values of c, the MSA quickly converges to the exact solution that is shown by the dashed line. The success of the MSA hinges on having a small parameter: for larger values of c, the convergence is slower. Table 6.2 lists the first several iterations for $a =$

12. If you are familiar with the Taylor series, please compare equations (6.27) with the Taylor expansion of the exact solution with respect to c that is truncated after the linear, quadratic, and cubic terms:

$$
\begin{aligned}
\tilde{x}_1 &= x_0 + \frac{dx}{dc}c = -\frac{c}{b}, \\
\tilde{x}_2 &= x_0 + \frac{dx}{dc}c + \frac{1}{2!} \cdot \frac{d^2x}{dc^2}c^2 = -\frac{c}{b} - \frac{ac^2}{b^3}, \\
\tilde{x}_3 &= x_0 + \frac{dx}{dc}c + \frac{1}{2!} \cdot \frac{d^2x}{dc^2}c^2 + \frac{1}{3!} \cdot \frac{d^3x}{dc^3}c^3 = -\frac{c}{b} - \frac{ac^2}{b^3} - \frac{2a^2c^3}{b^5}.
\end{aligned}
$$

For the truncation after the linear and the quadratic terms, the MSA yields the exact match with the Taylor series. For the truncation after the cubic term, the MSA matches the first three terms but adds another smaller fourth-order term.

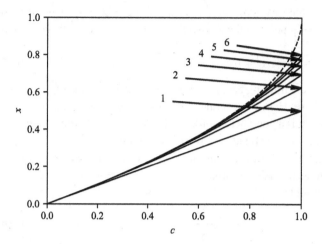

Figure 6.5
The quadratic equation: successive approximations

$1; b = -2; c = 0.1$. The exact solution for the root with a smaller absolute value is given by the quadratic formula; it matches the seventh MSA iteration for nine decimal places:

$$x = \frac{-b - \sqrt{b^2 - 4ac}}{2a} \approx 0.051316702. \tag{6.28}$$

Note that in this case the role of the small parameter is played by ax^2. Because of the hierarchy of scalings (see section 4.2), the contribution from ax^2 is expected to be small for the purpose of computing x in the left-hand side; therefore, this quadratic term can be used in lieu of a small parameter.

Table 6.2
Convergence for the quadratic equation

Iteration	x
0	0.0
1	0.05
2	0.05125
3	0.051313281
4	0.051316526
5	0.051316693
6	0.051316701
7	0.051316702

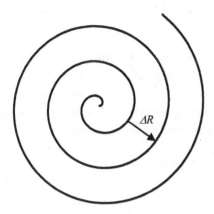

Figure 6.6
Archimedes's spiral: successive approximations

6.6 Archimedes's Spiral

The radius of Archimedes's spiral increases linearly with the turn angle (figure 6.6). Mathematically, this is expressed in polar coordinates as

$$r = a\theta. \tag{6.29}$$

The arc length of the spiral is given by the following formula:

$$L = \frac{\Delta R}{4\pi} \left(\theta \sqrt{1 + \theta^2} + \ln \left(\theta + \sqrt{1 + \theta^2} \right) \right), \tag{6.30}$$

where $\Delta R = 2\pi a$ is the distance between the adjacent loops and θ is the total turn angle.

Suppose we have a thread of length L and we want to embroider a piece of clothing with Archimedes's spiral that has the distance ΔR between the adjacent loops. What will the total turn angle for the spiral be? All attempts to solve equation (6.30) for θ using standard methods fail. Yet if we know that the spiral has multiple loops, this problem can be solved using the MSA, which becomes indispensable here.

The right-hand side of equation (6.30) contains the sum of two terms, $\theta \sqrt{1 + \theta^2}$ and $\ln \left(\theta + \sqrt{1 + \theta^2} \right)$. We assume that the spiral has multiple loops; this means that $\theta \gg 1$. In this case, the first term scales approximately as the power law $\left(\sim \theta^2 \right)$, and the second term scales logarithmically. From table 4.1 in section 4.2 we conclude that for large angles we must have $\ln\left(\theta + \sqrt{1+\theta^2}\right) \ll \theta\sqrt{1+\theta^2}$. This observation gives us a small term in the equation and a basis to use the MSA.

Key Point

Sometimes a problem does not have a small parameter. A scaling may produce a small term in the equation that can serve in lieu of a small parameter.

The next step is to solve for θ while temporarily ignoring (but not dropping outright) its presence in the smaller logarithmic term. From equation (6.30) we get

$$\theta\sqrt{1+\theta^2} = \frac{4\pi L}{\Delta R} - \ln\left(\theta + \sqrt{1+\theta^2}\right). \tag{6.31}$$

Squaring this equation produces

$$\theta^2\left(1+\theta^2\right) = \left(\frac{4\pi L}{\Delta R} - \ln\left(\theta + \sqrt{1+\theta^2}\right)\right)^2. \tag{6.32}$$

This is a quadratic equation for θ^2:

$$\theta^4 + \theta^2 - \left(\frac{4\pi L}{\Delta R} - \ln\left(\theta + \sqrt{1+\theta^2}\right)\right)^2 = 0. \tag{6.33}$$

The quadratic formula yields

$$\theta^2 = \frac{-1 + \sqrt{1 + 4\left(\frac{4\pi L}{\Delta R} - \ln\left(\theta + \sqrt{1+\theta^2}\right)\right)^2}}{2}, \tag{6.34}$$

where we retained only the positive root of the quadratic equation. Finally, we compute the square root to get

$$\theta = \sqrt{\frac{-1 + \sqrt{1 + 4\left(\frac{4\pi L}{\Delta R} - \ln\left(\theta + \sqrt{1+\theta^2}\right)\right)^2}}{2}}, \tag{6.35}$$

where we again retained only the positive root.

This equation definitely does not look simple, but it has a major advantage: we know that the occurrence of θ in the right-hand side is confined to a small logarithmic term. We can start by setting this term to zero, which would give us a zero-order approximation, and then work our way to a more accurate estimate.

In the zero-order approximation, we get

$$\theta_0 = \sqrt{\frac{-1 + \sqrt{1 + 4\left(\frac{4\pi L}{\Delta R}\right)^2}}{2}}. \tag{6.36}$$

The first-order approximation is computed if we substitute θ_0 in the right-hand side of equation (6.35) and so on:

$$\theta_1 = \sqrt{\dfrac{-1 + \sqrt{1 + 4\left(\frac{4\pi L}{\Delta R} - \ln\left(\theta_0 + \sqrt{1 + \theta_0^2}\right)\right)^2}}{2}},$$

$$\theta_2 = \sqrt{\dfrac{-1 + \sqrt{1 + 4\left(\frac{4\pi L}{\Delta R} - \ln\left(\theta_1 + \sqrt{1 + \theta_1^2}\right)\right)^2}}{2}},$$

(6.37)

$$\theta_3 = \cdots$$

Table 6.3
Convergence for the turn angle of Archimedes's spiral

Iteration	θ
0	11.187703109130538
1	11.047840163149468
2	11.048407209658428
3	11.04840489639464
4	11.048404905831351
5	11.048404905792856
6	11.048404905793012
7	11.048404905793012

Let us check if these iterations converge. Table 6.3 shows the results from several iterations for $L = 10$ and $\Delta R = 1$. The solution converges to $\theta \approx 11.048404905793012$. We can start stitching now.

Note again that the original equation (6.30) does not yield a closed-form solution for θ. For this problem, the MSA saves the day.

6.7 Designing Satellite Coverage

We are already familiar with the problem of designing satellite coverage. A satellite flies at altitude H above Earth and transmits a signal downward (figure 6.7). The antenna beam has a conical shape, and the angle from the center to the edge of the beam is α. Earth's radius is R. Section A.24 computes the length of the arc on Earth's surface from the center of the beam spot to its edge:

$$L = R\left(\sin^{-1}\left(\frac{R+H}{R}\sin\alpha\right) - \alpha\right).$$ (6.38)

In this section, we refine our solution: we account for the fact that Earth is not a perfect sphere. The shape of Earth is close to that of an ellipsoid that is slightly squeezed at the poles. Figure 6.7 shows the revised geometry of the problem, where the original spherical shape of Earth is shown in a dashed line, and the ellipsoidal shape is shown in a solid line (the effect of the ellipsoidal shape is exaggerated in the figure for display purposes). How does the nonspherical shape of Earth affect the coverage area?

Here we assume that the satellite is directly over the equator. To simplify this problem a bit, we limit the solution to finding angle β and will not extend it to finding the length of the arc L.

For a spherical Earth, we do obtain a closed-form solution for β in section A.24. For a nonspherical Earth, the problem becomes much more difficult. Our strategy is to follow generally the same path as in section A.24. The goal is to arrive at a final equation for β that will have the following general form:

$$\beta = F(\alpha, R_x, H, \mu \cdot g(\beta)),$$ (6.39)

where μ is a small parameter, R_x is the equatorial radius of Earth (now different from the polar radius), and where $F(), g()$ are some functions, to be determined later. Here a term (or terms) containing β in the right-hand side is multiplied by the small parameter μ. This would give us an opportunity to build a sequence of approximations. Every time we substitute a previous approximation for β in the right-hand side of this equation, the small multiplier μ will reduce the effect of any inaccuracy there. This will produce a more accurate next approximation.

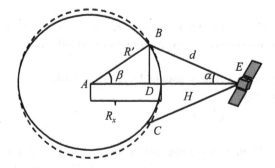

Figure 6.7
Designing satellite coverage: successive approximations

We identify a small parameter from the fact that the deviation of Earth's shape from spherical is quite small. As an added benefit, the zero-order approximation (assuming the limiting case $\mu = 0$) will default to the already solved problem for a spherical Earth.

We assume that the cross section of Earth is an ellipse and that its equation is given by

$$\frac{x^2}{R_x^2} + \frac{y^2}{R_y^2} = 1, \tag{6.40}$$

where R_x is the equatorial radius of Earth, and R_y is its polar radius. These radii are different; mathematically this is expressed as

$$R_y = R_x \cdot (1 - f). \tag{6.41}$$

Measurements show that $f \approx 1/298.257$. With this correction to the spherical model, we can modify the solution of the satellite coverage problem presented in section A.24. The new solution is as follows.

From the law of sines for triangle ABE in figure 6.7, we get

$$R' \sin \beta = d \sin \alpha, \tag{6.42}$$

where R' is the length of side AB. For nonspherical Earth, it is no longer equal to the equatorial radius R_x.

Next, the distance from the center of Earth to the satellite is the sum of the legs of two right triangles, ABD and DBE. These legs can be computed using the lengths of hypotenuses and the definition of cosine, but their sum is also equal to $R_x + H$. Therefore,

$$R' \cos \beta + d \cos \alpha = R_x + H. \tag{6.43}$$

Following the same logic as in section A.24, we get a formula that is similar to equation (6.38):

$$\beta = \sin^{-1} \left(\frac{R_x + H}{R'} \sin \alpha \right) - \alpha. \tag{6.44}$$

From figure 6.7 we see that R' is a function of β, so equation (6.44) is not a closed-form solution for β. To proceed further, we need to produce an explicit expression for $R'(\beta)$. If the origin of the coordinate system is located at the center of Earth, and if the coordinates of point B in figure 6.7 are denoted as x_B, y_B, then from the Pythagorean theorem we have

$$R'^2 = x_B^2 + y_B^2. \tag{6.45}$$

Since point x_B, y_B is located on the ellipse, these two variables are also linked by equation (6.40):

$$\frac{x_B^2}{R_x^2} + \frac{y_B^2}{R_y^2} = 1.$$ (6.46)

We transform this last equation as follows:

$$\frac{x_B^2}{R_x^2} + \frac{y_B^2}{R_x^2} + \frac{y_B^2}{R_y^2} - \frac{y_B^2}{R_x^2} = 1.$$ (6.47)

A multiplication by R_x^2 and collection of the terms yield

$$x_B^2 + y_B^2 + y_B^2 \frac{R_x^2 - R_y^2}{R_y^2} = R_x^2.$$ (6.48)

We substitute $x_B^2 + y_B^2$ from equation (6.45) and use $y_B = R' \sin\beta$ to get

$$R'^2 + R'^2 \frac{R_x^2 - R_y^2}{R_y^2} \sin^2\beta = R_x^2.$$ (6.49)

Next, we solve this for R':[13]

$$R' = \frac{R_x}{\sqrt{1 + \frac{R_x^2 - R_y^2}{R_y^2} \sin^2\beta}}.$$ (6.50)

Since $R_x \approx R_y$, the value of $\left(R_x^2 - R_y^2\right)/R_y^2$ is small. Using equation (6.41), we compute

$$\frac{R_x^2 - R_y^2}{R_y^2} = \frac{R_x^2 - R_x^2(1-f)^2}{R_x^2(1-f)^2} = \frac{2f - f^2}{(1-f)^2} \approx 0.006739.$$ (6.51)

We denote this small constant as μ. Next, we substitute R' in equation (6.44):

$$\beta = \sin^{-1}\left(\frac{R_x + H}{R_x}\sqrt{1 + \mu\sin^2\beta} \cdot \sin\alpha\right) - \alpha.$$ (6.52)

Here β is present in both sides of the equation. Strictly speaking, it is not a solution for β. However, the term containing β in the right-hand side is multiplied by a small constant $\mu \approx 0.006739$, as we have intended. This is a great opportunity to build a series of approximations. In the zero-order approximation, we assume that the perturbation term $\mu\sin^2\beta$ is zero. This defaults to the original spherical Earth problem (compare the equation below with equation (6.38), keeping in mind that $R_x = R$; $L = R\beta$ for a spherical Earth):

13. Since the derivation of equation (6.50) is cumbersome, it helps to check this result. It does pass the dimensional analysis test and satisfies limiting cases for $\beta = 0$ and $\beta = \pi/2$.

Table 6.4
Convergence for the satellite coverage problem

Iteration	β
0	0.765678
1	0.76802
2	0.768031
3	0.768031

$$\beta_0 = \sin^{-1}\left(\frac{R_x + H}{R_x}\sin\alpha\right) - \alpha. \tag{6.53}$$

Each next approximation is obtained by plugging the previous result in the right-hand side of equation (6.52):

$$\beta_1 = \sin^{-1}\left(\frac{R_x + H}{R_x}\sqrt{1 + \mu\sin^2\beta_0}\cdot\sin\alpha\right) - \alpha,$$

$$\beta_2 = \sin^{-1}\left(\frac{R_x + H}{R_x}\sqrt{1 + \mu\sin^2\beta_1}\cdot\sin\alpha\right) - \alpha, \tag{6.54}$$

$$\beta_3 = \cdots$$

Table 6.4 shows convergence for $R_x = 6{,}370$ km, $H = 20{,}000$ km, and $\alpha = 0.2$. The method performs excellently here, thanks to the diminutive value of μ, which produces a fast convergence. This is a formidable problem, with no obvious way to get a closed-form solution. Again, the MSA is instrumental for getting a result.

6.8 The Intersections between a Circle and a Parabola

A circle and a parabola are defined by the following equations (see figure 6.8):

$$x^2 + y^2 = R^2,$$
$$y = gx^2 + y_0. \tag{6.55}$$

In section A.6 we find the coordinates of intersections (if any) between these two curves:

$$y_{1,2} = \frac{-1 \pm \sqrt{1 + 4g(gR^2 + y_0)}}{2g},$$

$$y_{3,4} = y_{1,2},$$

$$x_{1,2} = \sqrt{R^2 - y_{1,2}^2}, \tag{6.56}$$

$$x_{3,4} = -x_{1,2},$$

where subscripts $1, 2$ in the notation $y_{1,2}$ correspond to the $\pm$ signs in the right-hand side. Solutions exist only if the expressions in the radicals are nonnegative. Depending on the signs of the expressions in the radicals, there can be zero, two, or four intersection points. In that problem, both curves are symmetric with respect to the y axis. This symmetry facilitates the solution.

What happens if the parabola is shifted left or right? Specifically, we want to solve the following system of equations:

$$x^2 + y^2 = R^2,$$
$$y = gx^2 + hx + y_0. \tag{6.57}$$

The linear term hx breaks the symmetry $x \leftrightarrow -x$, shifting the parabola left or right. If we imagine the parabola in figure 6.8 shifted to one side, we will see that the values of y at the intersections will no longer be pairwise equal. If there are four roots (for example, this happens if the parabola is squeezed in the x direction and its minimum is below the circle), all four values could be distinct. This hints at a fourth-order polynomial equation for the intersection points, which is indeed the case. Even though fourth-order polynomials do allow a closed-form solution, it is too bulky to be used in practice. How do we find the intersections in this case?

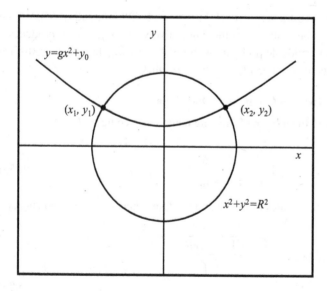

Figure 6.8
A circle and a parabola: successive approximations

> **Key Point**
>
> Try using more than one small parameter, separately or jointly. The MSA may provide insights in each case, showing different facets of the problem.

The MSA does not yield a general solution, but it helps in two particular cases, which we consider in turn.

6.8.1 A Small Linear Term

Here we consider the case of a small lateral shift of the parabola, which roughly corresponds to "small" values of h. The precise criterion for the small parameter will be clear from the solution.

The zero value of h reduces the problem to the already known one. We solve the new problem, temporarily ignoring the presence of the term hx in the right-hand side. We substitute y from the second equation of system (6.57) in the first equation:

$$x^2 + (gx^2 + hx + y_0)^2 = R^2. \tag{6.58}$$

After expanding the square and collecting the terms, we do get a fourth-order equation for x, as expected:

$$g^2 x^4 + 2ghx^3 + (2gy_0 + h^2 + 1)x^2 + 2hy_0 x + y_0^2 - R^2 = 0. \tag{6.59}$$

Next, we pile up all the terms containing h with the terms that do not contain x. This is in line with our strategy to temporarily ignore the presence of x in any small terms:

$$g^2 x^4 + (2gy_0 + 1)x^2 + \left(y_0^2 - R^2 + 2ghx^3 + h^2 x^2 + 2hy_0 x\right) = 0. \tag{6.60}$$

The solution for x^2 is given by the quadratic formula:

$$x^2 = \frac{-(2gy_0 + 1) \pm \sqrt{(2gy_0 + 1)^2 - 4g^2\left(y_0^2 - R^2 + 2ghx^3 + h^2 x^2 + 2hy_0 x\right)}}{2g^2}. \tag{6.61}$$

Finally, the solution for the intersection points is obtained by computing the square root of equation (6.61). Depending on the signs of the expressions under the square roots, there can be zero, two, or four intersection points:

$$x = \pm\sqrt{\frac{-(2gy_0 + 1) \pm \sqrt{(2gy_0 + 1)^2 - 4g^2\left(y_0^2 - R^2 + 2ghx^3 + h^2 x^2 + 2hy_0 x\right)}}{2g^2}}. \tag{6.62}$$

Just as in the previous examples, this is not a solution in the strict sense of the word because x appears in both sides of the equation. However, in the right-hand side x is always multiplied by small parameters h or h^2. This creates favorable conditions for applying the MSA. Just as before, we compute a zero-order approximation by setting $h = 0$:

$$x_0 = \pm \sqrt{\frac{-(2gy_0 + 1) \pm \sqrt{(2gy_0 + 1)^2 - 4g^2\left(y_0^2 - R^2\right)}}{2g^2}}. \tag{6.63}$$

Each subsequent approximation is computed by plugging the value of the previous approximation in the right-hand side of equation (6.62). Table 6.5 shows numerical values for the first few approximations. We used $R = 2, g = 1, y_0 = 1$, and $h = 0.01$. In this case, there are only two intersections between the circle and the parabola; both are shown in table 6.5.

Table 6.5
Intersection of a circle and a parabola (a small linear term)

Iteration	x_+	x_-
0	0.8895436175241324	−0.8895436175241324
1	0.8856103256735973	−0.8934284546302758
2	0.8856431077579379	−0.8934603843942568
3	0.8856428353860107	−0.8934606476328467
4	0.8856428376490879	−0.8934606498031196

Here subscripts $+, -$ correspond to the $\pm$ signs[14] for the outer square root in solution (6.62). The approximations quickly converge. Note that the two converged solutions for x are no longer symmetric with respect to $x \leftrightarrow -x$ because the linear term broke that symmetry. Only the zero-order approximation remains symmetric because it corresponds to the limiting case $h \rightarrow 0$, when the parabola is not shifted laterally and the symmetry is retained.

With the solution in hand, we can now address the issue of "smallness" for h. Suppose that all distances in the original equations are measured in meters. From the second equation in (6.57) we can conclude that h must be dimensionless. Therefore, the first impression is that the requirement for h to be small can be plainly expressed as $|h| \ll 1$. For example, let us look at our final result (6.62) and note that $|x| \leq R$ there, as the intersections must be located on the circle. We expect that if $|h| \ll 1$, then terms $2ghx^3, h^2x^2$, and $2hy_0x$ will not alter the expression in the second (internal) square root, which is dominated by larger

---

14. Note that to get successive approximations for x_+, we must use previous approximations for that very root in the right-hand side. Same is true for x_-.

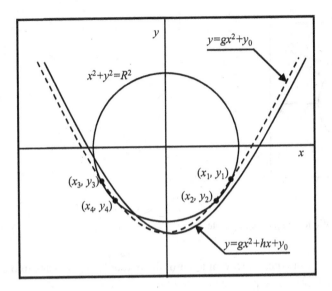

Figure 6.9
Criteria for h for the circle and parabola problem

terms, such as R^2. For many sets of parameter values this is indeed correct, and the criterion $|h| \ll 1$ is sufficient.

However, the situation becomes more complex when the expression under either square root is close to zero. In that case, even a small perturbation from h may change the result drastically. This is illustrated in figure 6.9. The dashed line shows the symmetric case, when $h = 0$; it produces four intersections. A slight shift ($h \neq 0$) of the parabola (shown in the solid line) causes it to miss the circle on the right. Even though $|h|$ may be small here, its effect may be large.

To summarize, criterion $|h| \ll 1$ works for many values of parameters, with the exception of some special cases. This illustrates the occasionally complex criteria for the validity of using a small parameter. In each problem, we must not just look for a plausible inequality, such as $|h| \ll 1$, but must analyze the impact that a small parameter has. We already saw another example of this in section 6.4, where MSA accuracy deteriorates for $a \approx b$, even if parameter d is nominally small.

6.8.2 A Small Quadratic Term

For the intersections of a circle and a parabola, another manageable case deals with a small value of the *quadratic* term in the second equation of (6.57). We start from a limiting case when the quadratic term is zero ($g = 0$) and proceed to a nonzero but small value of g.

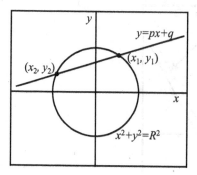

Figure 6.10
A small quadratic term in the circle and parabola problem

If we set the coefficient g for the quadratic term to zero, the parabola degenerates into a straight line, and we get the following equations:

$$x^2 + y^2 = R^2,$$
$$y = hx + y_0. \tag{6.64}$$

This is mathematically equivalent to the problem solved in section A.3, described by the following equations (see figure 6.10):

$$x^2 + y^2 = R^2,$$
$$y = px + q. \tag{6.65}$$

The intersections (if any) between a circle and a straight line are given by

$$x_\pm = \frac{-pq \pm \sqrt{(1 + p^2)R^2 - q^2}}{1 + p^2}. \tag{6.66}$$

These roots exist if the discriminant is nonnegative: $(1 + p^2)R^2 - q^2 \geq 0$. To get a solution of equation (6.64), we just have to set $p = h$; $q = y_0$ in equation (6.66). Therefore, the circle and straight line problem is a limiting case for the circle and parabola problem where $g \to 0$. If the quadratic term is nonzero but small, we can construct successive approximations to solve for x.

As in the previous examples, we use the solution for the limiting case $g = 0$ as a template and pay no attention to the presence of x in the right-hand side as long as it is confined to small terms. We use system (6.57) and substitute y from the second equation in the first one to get

$$x^2 + \left(gx^2 + hx + y_0\right)^2 = R^2, \tag{6.67}$$

where gx^2 is small. Expanding the expression in the parentheses in the last equation produces a fourth-order polynomial for x, but higher-order terms are small and are of no concern to us at this stage; we lump them with the free term. We get the same terms as in equation (6.60), but now the small terms are those containing g and not those containing h:

$$(1 + h^2)x^2 + 2hy_0x + (y_0^2 - R^2 + g^2x^4 + 2ghx^3 + 2gy_0x^2) = 0. \qquad (6.68)$$

Roots for x are found from the quadratic formula:

$$x_{\pm} = \frac{-hy_0 \pm \sqrt{h^2y_0^2 - (1 + h^2)\left(y_0^2 - R^2 + g^2x^4 + 2ghx^3 + 2gy_0x^2\right)}}{1 + h^2}. \qquad (6.69)$$

These roots exist if the discriminant is nonnegative. If $g = 0$, which corresponds to the true circle and straight line problem, the discriminant can be simplified a bit (see section A.3), but for $g \neq 0$ it remains bulky.

We have achieved our intermediate goal: even though x is present in the right-hand side of the last equation, it is always multiplied by small parameters g or g^2 there. This creates a basis for applying the MSA. We compute a zero-order approximation assuming $g = 0$ and then compute each next approximation by substituting the previous one in the right-hand side of equation (6.69):

$$x_0 = \frac{-hy_0 \pm \sqrt{h^2y_0^2 - (1 + h^2)\left(y_0^2 - R^2\right)}}{1 + h^2},$$

$$x_1 = \frac{-hy_0 \pm \sqrt{h^2y_0^2 - (1 + h^2)\left(y_0^2 - R^2 + g^2x_0^4 + 2ghx_0^3 + 2gy_0x_0^2\right)}}{1 + h^2}, \qquad (6.70)$$

$$x_2 = \frac{-hy_0 \pm \sqrt{h^2y_0^2 - (1 + h^2)\left(y_0^2 - R^2 + g^2x_1^4 + 2ghx_1^3 + 2gy_0x_1^2\right)}}{1 + h^2},$$

$$x_3 = \cdots$$

Table 6.6 shows numerical values for the first few approximations for roots x_+, x_-, where subscripts $+, -$ correspond to the $\pm$ signs in the solution.[15] We used $R = 2, g = 0.01, y_0 = 1$, and $h = 1$. Again, the approximations quickly converge.

---

15. Just as in the analysis of the small linear term above, we must consistently use x_+ for getting the next approximation for x_+, and x_- for getting the next approximation for x_-.

Table 6.6

Intersection of a circle and a parabola (a small quadratic term)

Iteration	x_+	x_-
0	0.8228756555322954	−1.8228756555322954
1	0.8181934381182916	−1.8329632626977677
2	0.8182586503368610	−1.8331995806206756
3	0.8182577460804985	−1.8332051563893987
4	0.8182577586200173	−1.8332052879680272

For this tough problem, the MSA leverages results from two simpler problems, which serve as limiting cases for the problem in hand. While the MSA does not provide us with a general solution, it is able to make inroads into it from two different starting points. Often this is the path one takes for dealing with difficult mathematical problems.

6.9   Summary

The previous chapters of this book presented tools to analyze, validate, and check an existing solution. This chapter is different: it presents a powerful method to *solve* mathematical problems. For many problems, this method works when all other approaches fail.

A mathematical purist may frown on the MSA. It involves making multiple iterations, and none of these iterations produces an exact result. Despite appearance, the same can be said about almost any so-called exact, closed-form solution if it uses such constants as π, or where the result is expressed through transcendental functions, for example, $\sin(x)$ or e^x. Indeed, the value of π is also computed using an iterative procedure, as are values of many functions. Having a convenient notation for such an iterative procedure (as it is the case for π or $\sin(x)$) does not change things fundamentally.

At the same time, the MSA is not a purely numerical technique. It can be used to gain an approximate analytical expression for the solution. This is important because such an expression can be dissected in many ways that we have studied above: we can check such elements as units, limiting cases, symmetries, and scaling.

The MSA has been widely used in physics. In the course of investigation researchers usually try to capture most important phenomena first and then extend the theory to weaker, less important effects. This strategy creates perfect conditions for applying MSA, making this method common for solving difficult problems. Because of its popularity, MSA comes in several flavors and may be called by different names. Sometimes, it is little more than a disguised Taylor series that you may know from calculus. This is not always the case: one can construct an example of an MSA when calculus is not applicable at all because the functions are not differentiable. This chapter presents an intuitive basis for one particular flavor of this method, which does not require knowledge of the Taylor series or other calculus topics.

In this implementation, we use the MSA to treat a problem that has a solvable limiting case where a parameter equals zero. For nonzero but small[16] values of this parameter, the solution of the full problem should not differ much from the limiting case. We follow the logic for the limiting case to solve for the unknown. For the limiting case, this produces an equation that has the unknown in the left-hand side but not in the right-hand side. For the full problem (when the small parameter is nonzero), the unknown remains present in the right-hand side, which is a complication. However, in the right-hand side the unknown is combined with a small parameter, which is key to applying the MSA.

Next, we build a sequence of approximations to the exact solution, each time plugging the previous approximation in the right-hand side. Often these approximations converge to the true solution. The smaller the small parameter, the better the speed of convergence.

While this method is powerful, it is not a panacea for difficult mathematical problems. A small parameter may not exist; even if there is a small parameter, the iterations may not converge. Still, the MSA has been a workhorse for countless applications in science and engineering.

This is the last chapter to introduce new material. The next two chapters present case studies. In each, we will take a real, practically important application and see how the various tools from this book apply to it. We will dissect and clarify mathematical equations to yield a better understanding of the problem and to produce new insights into it. These chapters will serve as examples and models for applying tools from this book to any problems that you may encounter in the future.

---

16. A reminder: a small value of a parameter makes sense only if that parameter is dimensionless.

Exercises

Problems with an asterisk present an extra challenge.

1. Section A.31 of this book presents a formula for mortgage payments. Given the initial loan amount D, the interest rate r, and the duration of the loan T, the annual payment rate on the mortgage is

$$p = \frac{rDe^{rT}}{e^{rT} - 1}. \tag{6.71}$$

Suppose that a customer has selected a house to buy and that she has a limited and known budget for paying the mortgage. What is the maximum interest rate that will keep the payments under budget?

The mortgage payment formula does not yield a closed-form solution for the interest rate r. Solve it for r using the MSA. Rewrite equation (6.71) as follows:

$$r = \frac{p}{D}(1 - e^{-rT}). \tag{6.72}$$

Assume that e^{-rT} serves as a small parameter here: any inaccuracy in the interest rate would be dampened by the exponent, yielding a more accurate value for r in the next approximation. Use the initial approximation $r_0 = 0.04$/year, the term of the mortgage $T = 30$ years, mortgage amount $D = \$230,000$, and a mortgage budget of $\$14,000$/year to estimate the maximum acceptable interest rate.

2. For each of the following equations, substitute numerical results for the zero-order approximation and for the highest computed order from the corresponding table of results. Draw conclusions on the accuracy of the MSA in each case.

    a) For equation (6.30), substitute values from table 6.3.

    b) For equation (6.52), substitute values from table 6.4.

    c) For equation (6.58), substitute values from table 6.5.

    d) For equation (6.67), substitute values from table 6.6.

3. Consider the following cubic equation:

$$(x - a)(x - b)(x - c) = d. \tag{6.73}$$

    a) What are the criteria for a, b, c, d for the method to be valid?

    b) For $a = 1, b = 2, c = 3$, and $d = 0.01$, estimate all three roots using the MSA. Produce both analytical formulas and numerical values for the first-order approximation.

    c) Check the numerical results by substituting them in the original equation (6.73).

4. The equation

$$\sin x = a \tag{6.74}$$

has one solution on the interval $0 \leq x \leq \pi/2$ for $0 \leq a \leq 1$:

$$x = \sin^{-1} a. \tag{6.75}$$

a) Use the MSA to find approximate formulas for the solutions of a modified equation:

$$\sin x = a + bx, \tag{6.76}$$

where b is small and $0 \leq x \leq \pi/2$. Limit your analysis to the first order.

b) From the formula for the first-order approximation, find a condition for a and b when the solution exists. (Since this condition is found from an approximate solution, it is also approximate.)

c) Plot both sides of equation (6.76) for $a = 0.8, b = 0.1$. Explain the condition on a and b that you found in task 4b.

5. A radar measures a range (distance) to the object it is tracking. Assume there are two radars that detect a sea vessel at ranges R_1 and R_2. Respective coordinates of the radars are $x_1 = 0; y_1 = 0$ and $x_2 = D; y_2 = 0$ (see figure 6.11). Section A.25 computes the coordinates of a sea vessel that is detected by the radars:

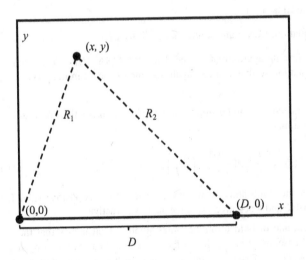

Figure 6.11
Approximations for detecting a vessel by two radars

$$x = \frac{D^2 + R_1^2 - R_2^2}{2D},$$

$$y = \pm \sqrt{R_1^2 - x^2}.$$

(6.77)

Though never mentioned explicitly, the solution in section A.25 assumes that Earth is flat. This is an approximation. Using dimensional analysis, specify a relevant small parameter or parameters for this approximation to be valid.

6. You are presented with the following equation for x:

$$e^{-x} = \tan x.$$

(6.78)

Assume that $x \gg 1$.

a) Plot both sides of this equation. How many solutions does it have?

b) Using a zero-order approximation $e^{-x} \approx 0$, isolate three different solutions $x = n\pi$, where $n = 1, 2$, and 5, and compute first-order approximations for these three solutions. How do they differ from the zero-order approximations? Is there a pattern? If yes, why?

7*. In this problem, you are again dealing with equation

$$e^{-x} = \tan x.$$

(6.79)

Now we assume that $x < 0$ and $|x| \gg 1$.

a) Plot both sides of this equation. How many solutions does it have?

b) Select one solution and find its approximate value using the MSA. (Hint: You need to transform the original equation so that you can again use the exponent in lieu of a small parameter.)

8. Problem 23 in chapter 4 introduced the Lennard–Jones model for atom interaction. It describes the force between two neutral atoms:

$$F(r) = \frac{24\epsilon}{\sigma}\left(2\left(\frac{\sigma}{r}\right)^{13} - \left(\frac{\sigma}{r}\right)^{7}\right),$$

(6.80)

where r is the distance between the atoms and σ, ϵ are positive parameters. A positive value for the force means that it is repulsive and a negative value means that it is attractive.

At large distances the magnitude of both terms becomes small, which means that the attraction force is weak. In that case, an atom on the surface of a liquid can break away from the surface because it is no longer constrained by the attraction force from other atoms. This process is known as evaporation.

a) Observe that both terms in the Lennard–Jones model exhibit power law scaling. Which term has a smaller magnitude for large values of r?

b) Assume that an atom has a high chance of leaving the surface of a liquid if $F(r) \geq -0.24\epsilon/\sigma$. Leverage the small value of one of the terms to find distances r at which this condition is true.

9. Among other requirements, a driverless car must quickly respond to emergency vehicles' sirens. In a driving range test, a siren is transmitted starting exactly at time $t_s = 0$ from a point with coordinates x_s, y_s. A test vehicle is moving in a circular loop. Its coordinates x_v, y_v are given by

$$x_v(t) = R \cos \frac{Vt}{R},$$
$$y_v(t) = R \sin \frac{Vt}{R}, \tag{6.81}$$

where R is the radius of the circular loop and V is the vehicle speed. To measure the latency of the vehicle's response, we need to compute the time T of the signal's arrival at the vehicle location. This time is delayed from $t_s = 0$ by the time needed for the sound to travel from the signal source to the vehicle. If the speed of sound is V_s, the signal travel time is given by

$$T = \frac{1}{V_s} \sqrt{(x_s - x_v(T))^2 + (y_s - y_v(T))^2}. \tag{6.82}$$

Substitution of $x_v(T), y_v(T)$ yields an equation for the time of the arrival of the signal:

$$T = \frac{1}{V_s} \sqrt{\left(x_s - R \cos \frac{VT}{R}\right)^2 + \left(y_s - R \sin \frac{VT}{R}\right)^2}. \tag{6.83}$$

This is a transcendental equation for T that cannot be solved by conventional methods. With this mathematical model in hand, do the following (note that cars do not drive close to supersonic speeds, and therefore $V \ll V_s$):

a) Using the MSA, produce formulas for the zero-order (which corresponds to $V_s \to \infty$), first-order, and second-order approximations for T.

b) Compute numerical values for the first- and second-order approximations using $R = 100$ m, $V_s = 340$ m/s, $V = 20$ m/s, and for the following two cases:

i. $x_s = 300$ m; $y_s = 100$ m

ii. $x_s = 100$ m; $y_s = 300$ m

c) What is the difference between the first- and second-order approximations for each location of the signal source? For one location, the first- and second-order approximations are closer than for the other location. Why?

10. This problem shows that a wrong application of the MSA may produce a diverging solution.

In chapter 7 we consider rare random events. In particular, we deal with the following formula for the probability of rare random events (see equation (7.11) in section 7.6):

$$\tilde{P}_n(x) \approx \frac{\sigma}{(x-\mu)\sqrt{2\pi}}e^{-\frac{(x-\mu)^2}{2\sigma^2}}. \tag{6.84}$$

If the events in question are rare, the probability of their occurrence is small: $\tilde{P}_n(x) \ll 1$. For brevity, we denote $y = (x-\mu)/\sigma$ to get

$$\tilde{P}_n(y) \approx \frac{1}{y\sqrt{2\pi}}e^{-\frac{y^2}{2}}. \tag{6.85}$$

For many applications, it is important to solve the last equation for y. Assume probability $\tilde{P}_n = 10^{-7}$, which corresponds to truly rare events.

a) Rewrite equation (6.85) as

$$y = \frac{1}{\tilde{P}_n\sqrt{2\pi}}e^{-\frac{y^2}{2}}. \tag{6.86}$$

Compute three MSA iterations using $y_0 = 5$.

b) Solving for y in the exponent of equation (6.85) produces

$$y = \sqrt{-2\left(\ln\left(\sqrt{2\pi}\tilde{P}_n\right) + \ln y\right)}. \tag{6.87}$$

Compute three MSA iterations using $y_0 = 5$.

c) Which approach converges, and which does not?

11. Solve the following equation for t, assuming that ω is a large parameter (rather than small):

$$\omega(t - t_0) = A\sin(ft). \tag{6.88}$$

Obtain analytical formulas for the zero- and first-order approximations.

12. The equation for x below is a combination of the equations for the sum of trigonometric functions (see section A.13) and for the product of two linear expressions (see section A.7):

$$f\sin x + g\cos x = c(x-a)(x-b). \tag{6.89}$$

Find analytical expressions for the zero- and first-order approximations for x for two cases:

a) Parameter c is small.

b) Parameter c is large.

c) If you roughly follow the MSA implementation in section 6.4 for the case of large values of c, the solution will break down if $a \approx b$. Provide a solution that is valid even if $a \approx b$.

13. You are given the following equation:

$$\sin e^{-x} = \sin x. \tag{6.90}$$

a) How many solutions does it have?

b) Select a solution with $x \gg 1$ and compute the zero- and first-order approximations for x.

14. The Lotka–Volterra equations describe the dynamics for two species, a prey (for example, rabbits) and a predator (for example, foxes). These equations have a stationary point, when the numbers of the prey animals and of the predators are constant, that is, do not vary over time. It is achieved when the decrease in the number of each species (for example, because of restricted food supply or being eaten by predators) is perfectly balanced with births. The stationary equations are

$$x(\alpha - \beta y) = 0,$$
$$-y(\gamma - \delta x) = 0, \tag{6.91}$$

where x is the number of prey, y is the number of predators, and α, β, γ, and δ are constants. In this model, the equilibrium is given by the solution of the above equations:[17]

$$y = \frac{\alpha}{\beta},$$
$$x = \frac{\gamma}{\delta}. \tag{6.92}$$

A researcher has come up with a more accurate version of the Lotka–Volterra equations, which modifies the stationary conditions as follows:

$$x(\alpha - \beta y) + \epsilon x^2 y^2 = 0,$$
$$-y(\gamma - \delta x) + \epsilon x^2 y^2 = 0. \tag{6.93}$$

For these modified stationary equations, find solutions for x and y, assuming that ϵ is small:

a) Use solution (6.92) as the zero-order approximation in the first equation of (6.93) to get the first-order approximation for y. Then use the zero-order approximation for x and the first-order approximation for y in the second equation of (6.93) to get the first-order approximation for x.

b) Use the zero-order approximation for x and y in both equations (6.93) simultaneously to get the first-order approximation for both variables.

---

17. The second stationary point is given by $x = 0; y = 0$. Zero numbers for both species at some time moment would obviously remain zero in the future, also making this solution stationary.

c) Use $\alpha = 1, \beta = 2, \gamma = 3, \delta = 4$, and $\epsilon = 10^{-2}$ to compute two different versions of first-order approximations, as explained in items 14a and 14b above.

d) Substitute the numerical values in equations (6.93) and determine which method yields better accuracy for the first-order approximation.

15. In section 2.4 we investigated limiting cases for the quadratic equation

$$ax^2 + bx + c = 0. \tag{6.94}$$

We have determined that for $c \rightarrow 0$ one of the roots approaches zero. Present this equation in the form:

$$x = -\frac{c}{ax + b}. \tag{6.95}$$

a) Use $a = 1, b = 2$, and $c = 10^{-2}$ and compute numerical values for MSA approximations of the orders 1 through 5. Use $x_0 = 0$ as the zero-order approximation.

b) Compute the exact value of x using the quadratic formula.

c) Compute the error in each of the approximations and plot it versus the approximation order number using the logarithmic scale for the vertical axis.

d) How does this error scale with the order number?

e) Section 6.5 presented a different way to solve the quadratic equation using the MSA. Use equations (6.27) to compute the numerical values for the first three approximations, and compare the results with those you obtained from applying MSA iterations to equation (6.95).

16. We again solve the quadratic equation:

$$ax^2 + bx + c = 0. \tag{6.96}$$

Rewrite this quadratic equation as

$$x = -\frac{ax^2 + c}{b}. \tag{6.97}$$

a) Use the quadratic formula to obtain both exact solutions for x when $c = 2, b = 1$, and $a = 10^{-3}$. Note that this case is different from the one we have investigated in section 6.5 and in exercise 15: we now assume a to be small instead of c.

b) Use $x_0 = 0$ and compute the first three approximations. Do they approach the exact solution?

c) Set $x_0 = -10^3$, which happens to be close to the other exact solution. Compute the first- and second-order approximations. Why do they not approach the exact solution, even though x^2 in the right-hand side is multiplied by a small parameter?

17. In section 6.1 we observed that the MSA works for the Achilles and the tortoise problem only if $\left|\frac{V_T}{V_A}\right| < 1$. Reconcile this inequality with the criterion for MSA convergence that we formulated in section 6.3.

7 Tying It All Together: The Probability of Catastrophic Events

This chapter presents a case study. We will explore how all the tools from the previous chapters work together on one important real-life problem: modeling rare but potentially catastrophic events.

Many phenomena show some combination of deterministic and random behavior. The former may predict a general trend or the most common outcome, and the latter is responsible for the deviations from it. As a rule, smaller deviations are more common than larger ones. Here we talk about rare, large deviations from the expected outcome (as opposed to the more common small deviations). For example:

1. Large earthquakes are rare, and small ones are more frequent. Large earthquakes are of primary concern to architects, city planners, and emergency services because they cause a loss of human lives and produce property damage.

2. An airplane should land approximately at the center line of the runway, but it may be off by a small amount. The pilot uses the navigation equipment to make sure the plane is not off by too much. A large navigation error, while extremely rare, may be a danger to the people on the plane.

3. A life insurance company bets on the typical longevity of the insured and on the expected investment returns for the securities it holds. A large, unpredicted deviation in either one may put the insurance policies at risk.

From these examples you see that rare events may have catastrophic consequences. This is why it is very important to estimate the probability of such events, especially for safety-critical applications. This probability would be a function of the magnitude of the event: the larger the event, the smaller the probability. Yet, even if the probability is small, it may be important to account for it.

7.1 Helpful Concepts from Probability Theory

If you studied probability theory, you may skip this section. It introduces a few concepts that are helpful for reading the material in this chapter.

7.1.1 What Is Probability?

There are multiple answers to this question, ranging from an axiomatic definition to vague qualitative explanations. For the purposes of this chapter, we will use a nonrigorous formulation based on the *frequency of a given outcome*. Suppose you have an event that can be repeated multiple times. Either because we lack knowledge about details of the processes that produce each event, or because these processes are intrinsically random, individual events yield different outcomes. If the total number of events is large (ideally, infinitely large) and a given outcome is observed in p fraction of all events, we say that the probability of that outcome is p. A common example of this is coin tossing. For a fair coin, the fraction of the cases when the coin lands heads up is close to 50 percent of all tosses.

This definition becomes difficult to apply when an outcome in question is rare, as is the case for the applications we will examine in this chapter. For such rare events, we would need a very large number of tries, which is impractical. In these situations we resort to mathematical models or computer simulations that will produce an estimate for the probability.

7.1.2 Random Variables

The outcomes of random events are often assigned numerical values. For a rolling die, we expect to get integer values from 1 to 6. A coin flip produces nonnumerical outcomes called *heads* and *tails*, but we are free to assign 0 to heads and 1 to tails.

We can introduce a variable whose value is given by individual outcomes of random events. Such a variable is called a *random variable*. These are a few examples of random variables:

1. In coin tossing experiments, the random variable may take values 0 and 1.
2. For a die, the random variable may take integer values from 1 to 6.
3. Above we talked about the navigation error during an airplane landing. Such an error (measured in meters) may potentially have any value. This means the corresponding random variable may take any real value.

7.1.3 The Expected Value and the Mean

We define the *expected value* $E(X)$ for random variable X as

$$E(X) = \sum_{j=1}^{n} p_j X_j, \tag{7.1}$$

where p_j is the probability of the random variable to have value X_j, and j enumerates all possible outcomes.

Suppose we make some number N of trials and record values X_k of a random variable, where the subscript $1 \leq k \leq N$ enumerates the trials. Let us compute the *mean* (also called the *population mean*) of all individual outcomes:

$$\bar{X} = \frac{1}{N} \sum_{k=1}^{N} X_k. \tag{7.2}$$

If the number of trials N is very large, and if we know the probability of each outcome, then this formula may be simplified. For example, consider three possible outcomes of a roulette game: black, red, and zero. Their respective probabilities are $p_b = 18/37$, $p_r = 18/37$, and $p_z = 1/37$.

Let us define a random variable by assigning numerical values $X_b = 1$ for the black, $X_r = 2$ for the red, and $X_z = 0$ for the zero. Consider now equation (7.2), where any term in the sum must be equal to one of the three values, $X_b, X_r,$ or X_z. For a very large number of trials, there will be approximately $p_b N$ of occurrences of $X_b = 1$ (black outcomes), $p_r N$ of occurrences of $X_r = 2$ (red outcomes), and $p_z N$ of occurrences of $X_z = 0$ (zero outcomes). If we group all occurrences by the three outcomes, equation (7.2) can be rewritten as

$$\bar{X} \approx \frac{1}{N}(p_b X_b N + p_r X_r N + p_z X_z N) = p_b X_b + p_r X_r + p_z X_z. \tag{7.3}$$

This equation has just three terms (instead of N), and its right-hand side fits the definition of the expected value of X in equation (7.1). This illustrates an important result: for a large number of trials, the population mean approaches the expected value.

7.1.4 Cumulative Probability Distribution

For a random variable X, the *cumulative probability distribution* is a function $P(x)$ that shows the probability that the random variable X has a value less than or equal to x. The *complementary cumulative probability distribution* flips the sign of the inequality in the definition: it is a function $\tilde{P}(x)$ that computes the probability that a random variable X has a value *greater* than x. In this chapter, we will deal with the complementary cumulative probability distribution function (CCPDF).

For example, consider the roulette game again. We have defined the random variable X that takes three values: 0, 1, and 2. For this random variable, we can define the complementary cumulative probability distribution function as follows:

1. Since X never takes a value greater than 2, the probability of $X > 2$ is zero. Therefore, $\tilde{P}(x) = 0$ for $x \geq 2$.

2. For any x in the interval $1 \leq x < 2$, there is only one possibility for the random variable X to exceed x: it happens when $X = 2$, which has the probability of 18/37. Therefore, $\tilde{P}(x) = 18/37$ for $1 \leq x < 2$.

3. For $0 \leq x < 1$, there are two possibilities for the random variable X to exceed x. It happens when $X = 1$ or when $X = 2$. The total chances of that compute to $(18/37) + (18/37) = 36/37$. Therefore, $\tilde{P}(x) = 36/37$ for $0 \leq x < 1$.

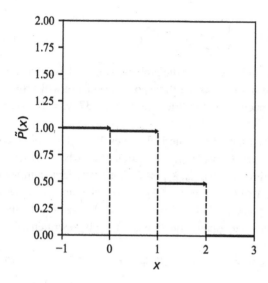

Figure 7.1
CCPDF for the roulette game

4. Finally, X is always nonnegative and therefore always exceeds any negative threshold. This means that for $x < 0$, the probability of $X > x$ is equal to 1, and $\tilde{P}(x) = 1$ for $x < 0$.

This function is plotted in figure 7.1.

Armed with these concepts, we are ready to deal with the problem of rare but catastrophic events.

7.2 Generalized Pareto Distribution

The modern theory of so-called extremal events[1] predicts that in many cases the probability of having a large event with a magnitude greater than x is given by the *generalized Pareto distribution* (GPD). The CCPDF in this case is as follows:

$$\tilde{P}(x) = \left(1 + \xi\frac{x - \mu}{a}\right)^{-\frac{1}{\xi}},\tag{7.4}$$

---

1. A monograph by Sornette (2006) presents a great overview of the topic.

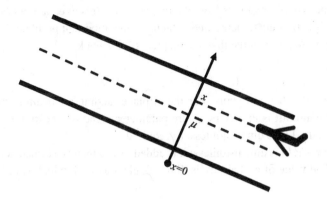

Figure 7.2
Airplane landing error

where $\tilde{P}(x)$ is the probability of a deviation from the mean[2] exceeding the value of x, and μ, ξ, and a are parameters. Values of these parameters are application specific and in practice may be estimated from available data. Below we assume that $\xi, a > 0$.

For example, let's consider again the example of an airplane that is landing on a runway (see figure 7.2). The transverse coordinate of the airplane position is denoted as x. In the middle of the runway, $x = \mu$. In reality, condition $x = \mu$ holds only approximately, and there is a small navigation error, putting the airplane slightly off the center line. In an unlikely case when several error sources conspire to form a large navigation error, cockpit gauges may still show that the airplane is within the acceptable bounds, but in reality the plane may be landing off the runway. For large values of $x - \mu$, the probability of such an event is given by equation (7.4), where x, μ, and a are measured in meters, and ξ is a dimensionless parameter.[3]

It is the job of aerospace engineers and data scientists to design the plane navigation system in such a way that this probability is negligibly small and that in practice it does not lead to catastrophic consequences.

2. The mean of a probability distribution (when it exists) is the expected value of the underlying random variable for that distribution. For a large number of trials, the population mean is approximately equal to the mean of the distribution, justifying the use of the latter term.

3. Equation (7.4) does not say anything about the probability of small, frequently occurring deviations, which may be described by completely different math. Quite often it is the ubiquitous bell curve.

Below we apply all of the analysis methods from the previous chapters to equation (7.4) and show how various tools contribute to better understanding this important problem. We follow the same sequence of topics as in the first six chapters of this book.

7.3 Units

Unit dimensionality is the first check to apply. For the airplane landing application, the transverse position x of the aircraft is in meters, as are parameter a and the location μ of the center line. The exponent $-1/\xi$ is dimensionless, as required.

Therefore, expression $\xi(x-\mu)/a$ is dimensionless; it is added to 1, which is a dimensionless numerical constant. The value of probability is dimensionless as well, which is to be expected.

The unit check works.

7.4 Limiting Cases

There are three limiting cases to consider in equation (7.4):

1. The first limiting case is given by the fact that larger deviations are ever more rare. Therefore, for large values of x, the probability of observing deviations from the mean that exceed the value of x must be small:

$$\tilde{P}(x \to \infty) \to 0. \tag{7.5}$$

   Indeed, for $x \to \infty$ and for positive values of ξ, we do get $\tilde{P}(x \to \infty) \to 0$ in equation (7.4). This limiting case holds.

2. The probability of getting a deviation from the mean $x - \mu$ of *any* size should be 1. This means that we should see $\tilde{P}(x = \mu) = 1$. We can check this by direct substitution, which does yield the desired result.

   As noted above, equation (7.4) does not apply to small deviations from the mean. Since $x = \mu$ removes a threshold for deviations from the mean, this case includes small deviations (as well as large ones), and this limiting case does not have to work. Even so, it works, which is an unexpected benefit.

3. In the third limiting case[4] we deal with limit $\xi \to 0$. We rewrite equation (7.4) as follows:

$$\tilde{P}(x) = \left[\left(1 + \xi \frac{x - \mu}{a} \right)^{\frac{a}{\xi(x-\mu)}} \right]^{-\frac{x-\mu}{a}}. \tag{7.6}$$

   The value of $\xi \frac{x-\mu}{a}$ is small, and the expression in the square brackets is related to the definition of number e:

---

4. If you are not familiar with calculus, you may skip this limiting case.

$$e = \lim_{n \to \infty} \left(1 + \frac{1}{n}\right)^n .$$ (7.7)

Therefore, for small values of ξ, the probability of a random variable exceeding x is given by

$$\tilde{P}(x) = e^{-\frac{x-\mu}{a}} .$$ (7.8)

This is known as the exponential distribution. Unlike many of the examples above, this limiting case does not serve as a check on a foreknown result. Yet, it helps to better understand a scaling behavior in section 7.6 below.

7.5 Symmetry

The important symmetry here is related to the space-shift invariance. We are already familiar with this invariance from exercise 21 in chapter 3.

A change in the value of μ shifts the entire GPD by a fixed amount. While seemingly trivial, this invariance is fundamental. It states that our choice of the origin of the coordinate system (corresponding to the zero value for x) is arbitrary, and the result must not depend on it.

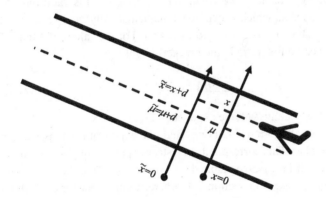

Figure 7.3
Airplane landing: shift invariance

Figure 7.3 illustrates this concept. Variables $\tilde{x}, \tilde{\mu}$ are shifted with respect to variables x, μ by amount d, attributed to a different definition for the $x = 0$ point. Even so, the total arrangement remains the same, and the difference $(x - \mu)$ remains invariant:

$$\tilde{x} - \tilde{\mu} = x - \mu. \tag{7.9}$$

As a result, the probability of the airplane landing by more than x meters off the track is independent of the coordinate origin, as expected.

The space-shift invariance is universal. The same is true for time-shift invariance, which we will explore in the next chapter.

7.6 Scaling

There are three scaling features in this problem; one of them has huge practical implications.

We start from a trivial one: the probability remains invariant if we replace $x \leftrightarrow \alpha x$; $a \leftrightarrow \alpha a$; and $\mu \leftrightarrow \alpha \mu$. This effectively decreases the amount of independent variables in play here. For example, if we obtain results for one particular value of x and for a range of values of a and μ, we can scale these results to cover other values of x without doing computations again.

To see the remaining scaling behaviors, we recall that equation (7.4) is valid only for large deviations of x from the mean, which is expressed mathematically as $x - \mu \gg a$ and (according to limiting case 1 above) corresponds to $\tilde{P}(x) \ll 1$. For the values of ξ that are not very small, we can neglect the first term in the brackets to get

$$\tilde{P}(x) \approx \left(\xi \frac{x - \mu}{a} \right)^{-\frac{1}{\xi}}. \tag{7.10}$$

For $(x - \mu)$, this is a power law scaling with the exponent $-1/\xi$.

What happens if ξ is very small ($\xi \ll 1$)? On one hand, the exponent in the power law becomes a large negative number. This corresponds to a steeper power law decrease in the value of $\tilde{P}(x)$ for large values of $(x - \mu)$ (see table 4.1 in section 4.2). On the other hand, this was the subject of limiting case 3 in section 7.4, where we saw that for $\xi \to 0$, the distribution reduces to the exponential one. To summarize: as the value of ξ decreases, we get steeper and steeper power distributions, and in the limit of $\xi \to 0$ the distribution is no longer a power law but an exponential one. This is intuitively consistent with the fact that an exponent beats *any* power law, just as stipulated in table 4.1. For many practical applications, ξ is finite and the distribution for large deviations from the mean remains a power law one.

For modest deviations from the mean, equation (7.4) is not applicable, and the distribution there can be different. Very commonly, for modest deviations from the mean there are good reasons to use the so-called normal distribution. The trouble starts when researchers apply the normal distribution to estimate the probability of rare events, where they are actually described by the GPD formula (7.4).

To appreciate the difference between the normal distribution and the GPD, we have here an approximate expression for the probability of a rare event if it were described by the normal distribution:[5]

$$\tilde{P}_n(x) \approx \frac{\sigma}{(x - \mu)\sqrt{2\pi}} e^{-\frac{(x-\mu)^2}{2\sigma^2}}, \tag{7.11}$$

where σ^2 is called the *variance*,[6] and where $\tilde{P}_n(x)$ is the probability of a rare event exceeding the value of x. In other words, it is the normal CCPDF for large values of x.

The scaling of $\tilde{P}_n(x)$ is exponential for $(x - \mu)^2$, which is extremely fast—even faster than the standard exponential scaling in table 4.1 in section 4.2. This is because we have nested exponential and quadratic functions here, and $\tilde{P}_n(x)$ goes down very quickly as a result. Therefore, large deviations from the mean in the case of the normal distribution are extremely rare. We have learned from table 4.1 that for negative exponents any power law will produce larger values than an exponential function for large enough values of x. For the problem in hand this translates into

$$\tilde{P}(x) \gg \tilde{P}_n(x) \tag{7.12}$$

for large enough values of $(x - \mu)$. This is true regardless of the value of σ!

Therefore, an indiscriminate application of the ubiquitous normal distribution (as opposed to using the GPD where appropriate) commonly underestimates the probability of rare events. As you see from the examples of rare events listed early in this chapter, this can have bad consequences: we may be lulled into falsely thinking that the chances of a large earthquake or of an air traffic accident are very low and neglect to design the systems that would protect us from such events.

This section shows that scaling can alert us to a potentially serious or even disastrous problem in applying an incorrect mathematical model. Below we quantify this important conclusion.

7.7 Order of Magnitude Estimates

First, we make an estimate for a value of x that yields a significant change in the probability value $\tilde{P}(x)$ in equation (7.4). Since the GPD possesses shift invariance, we can choose the origin of the x axis in such a way that $\mu = 0$; this simplifies the formulas a bit. The

---

5. This formula may seem different from the standard expression that you may find elsewhere. We used an approximation for large values of $(x - \mu)/\sigma$ to stay within the confines of algebraic functions and to avoid using the so-called error function. Note that it does not have that nice limiting case $\tilde{P}_n(x) \to 1$ for $x \to 0$, but as discussed above, satisfying that limiting case is not mandatory for a formula derived in the assumption of a large $x - \mu$.

6. Variance is defined in section A.32.

only remaining parameter measured in meters is a; this parameter defines the scale for x. Indeed, if we change x by an amount on the order of a, the value of $\tilde{P}(x)$ changes in a significant way. According to rule 8 in section 5.2, this significant change in $\tilde{P}(x)$ implies a corresponding significant change in x. Similar reasoning points to σ as the scale of a significant change for the normal distribution (7.11).

Thus, parameters a (for the GPD) and σ (for the normal distribution) determine the characteristic scales of x at which a significant change in the probability is achieved. These parameters define the widths of the respective distributions. If an engineer analyzes data for the roll-off of a distribution and makes a mistake by using a normal distribution model instead of the GPD, his estimate for the value of σ in the normal distribution may be on the order of the value of a in the true underlying GPD distribution. Hence, to estimate the effect of such a mistake in selecting the correct model, it makes sense to set $\sigma \sim a$ for an apples-to-apples comparison of these two distributions.

With this setup, we return to the last paragraph of the previous section, which prompts us to compare the GPD and the normal distribution quantitatively. Specifically, we compare the probabilities of rare events as given by equations (7.4) and (7.11).

To make equations (7.4) and (7.11) valid, we need to ensure that we deal with truly rare events. We already saw that this requires that $x \gg \sigma, a$ (note that μ is no longer a part of the criterion here because we have set it to zero). We consider the case when x is one order of magnitude greater than σ and a:

$$x = 10\sigma = 10a. \tag{7.13}$$

Then from equation (7.11) we get

$$\tilde{P}_n(x) \approx \frac{\sigma}{x\sqrt{2\pi}} e^{-\frac{x^2}{2\sigma^2}} \sim 10^{-23}. \tag{7.14}$$

This is a truly negligible value. If we trust this result, we can use it to design airplane landing systems, nuclear power stations, earthquake-proof buildings, and other safety-critical applications.

Let us compare this with the values given by equation (7.4). First, we need to select a value for parameter ξ, which is application specific. According to rule 9 in section 5.2, we can use a value on the order of 1. We have speculated that the normal distribution will go down faster than the GPD for any value of ξ. To make a stronger case than $\xi \sim 1$, we consider $\xi = 1/4$; in that case the probability $\tilde{P}(x)$ scales as x^{-4} for large values of x. We will see that even this relatively fast power law decrease in $\tilde{P}(x)$ will produce markedly alarming estimates.

For these parameter values, we get the following for the GPD:

$$\tilde{P}(x) = \left(1 + \xi\frac{x}{a}\right)^{-\frac{1}{\xi}} = 3.5^{-4} \sim 0.007. \tag{7.15}$$

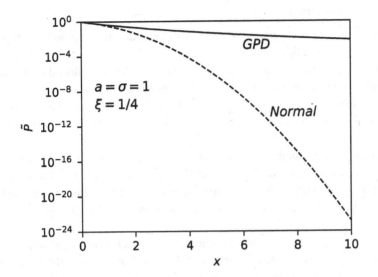

Figure 7.4
CCPDF for the GPD and the normal distribution

The difference between these two values is about 21 orders of magnitude! Using the normal distribution instead of the GPD for a safety-critical application would expose us to a very large risk.

Figure 7.4 shows plots of CCPDFs for the GPD and the normal distribution.[7] The difference between the GPD and the normal distribution is many orders of magnitude starting at moderate values of x.

Even if we deal with a faster exponent in the GPD, the gaping difference between these two results is not reduced much. For example, using $\xi = 0.1$ (which is a quite fast power law decrease with the exponent equal to 10) produces $\tilde{P}(x) \sim 10^{-3}$. A system with this margin of safety will fail one time per thousand uses, which is definitely not good.

Rough order of magnitude estimates in this section show the importance of choosing an adequate model to estimate the risk of failure for various systems. The difference is stark and cannot be ignored.

---

7. For the normal distribution we plot the probability that $|X| > x$ rather than $X > x$; this helps the probability values for both curves equal 1 at $x = 0$. We also use exact values instead of approximate formula (7.11).

7.8 Successive Approximations

The previous section established the importance of using the GPD where appropriate. However, a practical application of this mathematical model requires knowledge of three parameters: μ, ξ, and a. The only way to estimate these parameters is to collect data on the value of a random variable exceeding different thresholds (that is, corresponding to different values of x) and then to infer parameter values from those data. Here we explore a simpler version of this problem: we assume that $\mu = 0$ and that a has been estimated from a separate set of data.

Suppose we studied a large number of observations and came to a conclusion that for $\mu = 0$ and $x/a = 10$, the error in the airplane landing position exceeds the threshold x approximately 5 times per 1,000 cases, indicating that $\tilde{P}(x) = 5 \cdot 10^{-3}$. This is a very high level of risk, and we conclude that the threshold $x/a = 10$ is unacceptable. Suppose that to ensure safe airplane landings, we require $\tilde{P}(x) = 10^{-8}$.

To meet this stringent requirement, we must clamp down on the airplane navigation error, which can be achieved by using better instruments. This will drive down the value of a, increasing the ratio x/a and correspondingly decreasing $\tilde{P}(x)$ in equation (7.4). The important question is: how much better must the navigation accuracy be? Is it sufficient to increase the value of x/a by a factor of 100, or do we need more? To answer this question, we need to do two things:

1. Estimate the value of ξ.
2. Use equation (7.4) to solve for the value of x/a that yields the desired value of risk $\tilde{P}(x) = 10^{-8}$.

Since a closed-form solution for ξ is out of reach, we turn to the method of successive approximations (MSA). We solve for the exponent in equation (7.4) assuming $\mu = 0$ and transform the result as follows:

$$\xi = -\frac{\ln\left(1 + \xi\frac{x}{a}\right)}{\ln \tilde{P}(x)}. \tag{7.16}$$

From the data we know that $\tilde{P}(x) = 5 \cdot 10^{-3}$ for $x/a = 10$. We also use $\xi = 0.2$ as a zero-order approximation. Higher-order approximations are constructed as follows:

$$\xi_n = -\frac{\ln\left(1 + \xi_{n-1}\frac{x}{a}\right)}{\ln \tilde{P}(x)}. \tag{7.17}$$

Table 7.1 shows convergence to a solution for ξ.

Armed with the last value of ξ, we can determine the new threshold value. We raise both sides of equation (7.4) to power $-\xi$ and then solve the resulting equation for x/a:

$$\frac{x}{a} = \frac{1}{\xi}\left(\left(\tilde{P}(x)\right)^{-\xi} - 1\right). \tag{7.18}$$

Table 7.1

Approximations for parameter ξ

Iteration	ξ_n
0	0.2
1	0.207351166920353
2	0.211920256261329
3	0.214705405800963
4	0.216383186633123
5	0.217386734168391
6	0.217984455444685
7	0.218339565812304
8	0.218550223399560
9	0.218675078090012

For $\tilde{P}(x) = 10^{-8}$ and $\xi = 0.2187$ we get $x/a \approx 252$.

For example, assume that the width of the airplane runway is 80 m. Let us assume that this allows for a 25 m navigation error on each side, which means that x is set to 25 m. We have also assumed that a safe operation requires $\tilde{P}(x) = 10^{-8}$. According to the numerical estimate above, this leads to $x/a \approx 252$, and therefore we get $a \approx 0.1$ m. Section 7.7 argued that a is the scale of a significant change in x, which means that the probability for the navigation error to exceed x rolls off roughly at the scale of a, and that a defines the width of the GPD. Then the airplane navigation equipment must have the typical accuracy on the order of a, or about 0.1 m. Such accuracy would lead to a 25 m error about one time in 10^8 landings, ensuring a safe operation.

7.9 Summary

In this chapter we worked through a very important application. By our nature, we are prone to discount the consequences of rare events. Yet, for some of them, such as strong earthquakes or large errors in airplane navigation, the outcomes can be catastrophic. Accurate mathematical models for the probability of such events are literally lifesaving.

To design a safe system, engineers must perform many analyses, including some similar to those we have done in this chapter. Our treatment is grossly simplified; in practice, the calculations that go into ensuring safe operation of airplanes, nuclear power plants, and other safety-critical applications are much more comprehensive and more data driven. Yet, even within the constraints of a simplified analysis, the tools that we learned in the previous chapters have led us to select the GPD over the normal distribution, which makes a huge difference. We even obtained a rough estimate for the requirement for a typical navigation error. This illustrates the power and the importance of the tools presented in the first six chapters.

8 Tying It All Together: Tracking a GPS Satellite

This chapter presents another example of analyzing a real-life application. For this purpose, we will examine a mathematical model for a GPS receiver tracking a satellite. This problem is more complex than the one in chapter 7, mostly because of the more cumbersome equations. Unfortunately, high complexity is common and often is the price one has to pay for getting things right. For complex applications, the benefit of being able to check your solution or to get an additional insight becomes even more important.

Section 8.1 gives some (oversimplified) background information. It culminates with a rather bulky equation for tracking a signal by a GPS receiver. In the subsequent sections, we will apply various tools to analyze and dissect this equation.

8.1 Problem Setup

A GPS receiver continuously tracks signals from several GPS satellites. Satellite orbits are known with very high accuracy, and each satellite transmits a time-stamped signal. A receiver essentially measures the distance to each satellite by comparing the time of the arrival of that signal with the time stamp of its transmission. It then uses these distance measurements to solve for position. The primary algorithm that does signal tracking is called the phase-locked loop (PLL).

Each GPS satellite is moving; the receiver can also move or be stationary. If the distance between the satellite and the receiver changes, the algorithm must detect this change. Because of the nature of signal processing algorithms, the detection of a change in the distance is not perfectly accurate and is not instantaneous. An everyday analogy for this process is driving a car on a curvy road. As the road veers off to the left or to the right, the driver must react to keep the car on the road. His control of the car cannot be perfect and may have a small delay, but overall the path of the car follows the road reasonably accurately. Mathematically, the tracking is characterized by two functions:

1. The true change in the distance between the satellite and the receiver as a function of time, $u(t)$.
2. The measured change in distance, $y(t)$.

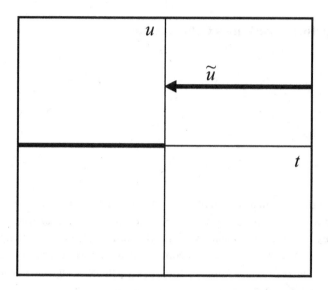

Figure 8.1
The step function

The goal is to design a GPS receiver where $y(t)$ faithfully approximates $u(t)$.

The mathematical model for the tracking process is quite complex and is beyond the scope of this book. Moreover, here we present a result for only one specific form of function $u(t)$, the step function (see figure 8.1), defined as

$$u(t) = \begin{cases} 0, & \text{if } x \le 0, \\ 1 & \text{if } x > 0. \end{cases} \tag{8.1}$$

In reality, the distance between a satellite and a GPS receiver cannot jump instantaneously by a finite amount, so equation (8.1) is an abstraction. Yet, it can be used as a building block to model a receiver response to *any* function, so this abstraction is quite useful. Figure 8.2 shows an example of how a smoothly varying function can be approximated as a stacked combination of small step functions.[1] This explains why characterizing a receiver's response to the step function is important.

If the change in the distance between the satellite and the GPS receiver is modeled as the step function, the receiver response is given by the following formula:[2]

---

1. The exact way of doing so is given by calculus; if you are familiar with it, you may see in figure 8.2 a resemblance to the Riemann sum concept.

2. A textbook by Misra and Enge (2010) discusses GPS tracking in depth.

$$y(t) = \left\{ 1 - \frac{e^{-2\pi\zeta f_N t}}{\sqrt{1 - \zeta^2}} \left[\sin\left(2\pi\zeta f_N \sqrt{1 - \zeta^2} \cdot t + \phi\right) \right.\right.$$
$$\left.\left. - 2\zeta \sin\left(2\pi\zeta f_N \sqrt{1 - \zeta^2} \cdot t\right) \right] \right\} u(t), \tag{8.2}$$

where ϕ is defined as

$$\phi = \tan^{-1} \frac{\sqrt{1 - \zeta^2}}{\zeta}, \tag{8.3}$$

and where ζ and f_N are design parameters of the receiver. The values of these parameters can be selected by engineers to optimize receiver performance. Parameter f_N is called the *loop bandwidth* and is measured in s^{-1}. Parameter ζ is dimensionless and is constrained to the 0 to 1 range:

$$0 < \zeta < 1. \tag{8.4}$$

As noted above, mathematical model (8.2) applies to the case when $u(t)$ is a step function. Since any realistic function can be approximated as a combination of small steps, equation (8.2) can be used to construct the receiver's response to virtually any input.

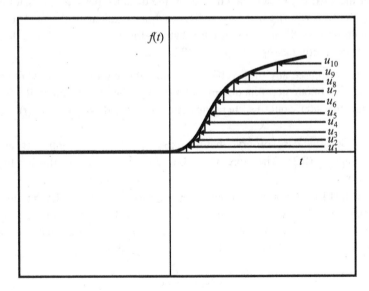

Figure 8.2
Step functions as building blocks for a general input

With this background in hand, we proceed to the analyses of equation (8.2).

8.2   Units

We start from analyzing units and dimensionality. This is the first check that should be applied to any formula, especially one that results from a lengthy and complex derivation, as is the case for equations (8.2) and (8.3).

Parameter ζ is dimensionless, as are $2\pi\zeta$, $\sqrt{1 - \zeta^2}$, and similar expressions. Therefore, the argument of $\tan^{-1}$ in the definition of ϕ is dimensionless, as it should be. Since ϕ is the value of the inverse tangent, it is also dimensionless.

The loop bandwidth f_N is measured in s^{-1}, and the time t is measured in s; therefore, the product $f_N t$ is dimensionless.

We conclude that the arguments of the sine and exponential functions in equation (8.2) are dimensionless, as expected. The value of the entire expression in the curly brackets is dimensionless as well, and the units for $y(t)$ are the same as the units for $u(t)$. Therefore, if the true change in the distance $u(t)$ between the satellite and the GPS receiver is measured in meters, the perceived value $y(t)$ is also measured in meters, which makes sense.

Equations (8.2) and (8.3) pass the unit check.

8.3   Limiting Cases

The design goal of the PLL is to track the changes in the distance between the satellite and the receiver. If the PLL algorithm in the receiver and our mathematical model for it are correct, we should observe three commonsense features for function $y(t)$. These three features constitute three limiting cases for equation (8.2):

1. The receiver cannot foresee any future changes in $u(t)$; it can only react to what has happened in the past. In our step-function model, $u(t) = 0$ when $t \leq 0$, and until $t = 0$, nothing indicates that $u(t)$ would be changing. Therefore, we must see that $y(t) = 0$ when $t \leq 0$.

   To verify this limiting case, observe that the expression for $y(t)$ contains $u(t)$ as a multiplier (see equation (8.2)). Therefore, $y(t) = 0$ whenever $u(t) = 0$, and this limiting case holds.

2. As noted above, the PLL cannot react instantaneously to a change in $u(t)$. Even though $u(t)$ is discontinuous at $t = 0$, the measured value needs a bit of time to ramp up. Therefore, we expect $y(t)$ to be continuous. At $t = 0$, the value of $y(t)$ will start varying from the zero point upward. Mathematically, this is expressed as follows:

$$y(t) \to 0 \quad \text{when} \quad t \to 0. \tag{8.5}$$

Qualitatively, this means that for small t, the value of $y(t)$ must be small.

Checking this limiting case is a bit more tricky. We start from inequality $0 < \zeta < 1$. This allows us to introduce a new variable ψ, such that $\zeta = \cos\psi$ and $\pi/2 > \psi > 0$. Then $\sqrt{1 - \zeta^2} = \sin\psi$. Equation (8.3) takes the form

$$\phi = \tan^{-1}\frac{\sin\psi}{\cos\psi} = \psi. \tag{8.6}$$

Turns out, our new variable ψ and the original variable ϕ are one and the same. We can now replace $\sqrt{1 - \zeta^2}$ with $\sin\phi$ in the denominator in equation (8.2):

$$y(t) = \left\{ 1 - \frac{e^{-2\pi\zeta f_N t}}{\sin\phi}\left[\sin\left(2\pi\zeta f_N \sqrt{1 - \zeta^2}\cdot t + \phi\right)\right.\right.$$
$$\left.\left. - 2\zeta\sin\left(2\pi\zeta f_N \sqrt{1 - \zeta^2}\cdot t\right)\right]\right\}u(t). \tag{8.7}$$

Let us now examine this last equation. If $t \to 0$, the argument of the second sine function in the square brackets approaches zero, and the sine approaches zero as well. The argument of the first sine function approaches ϕ. The argument of the exponent approaches zero. As a result, we get for $t \to 0$:

$$y(t) \to \left\{ 1 - \frac{1}{\sin\phi}\left[\sin\phi\right]\right\}u(t) = 0. \tag{8.8}$$

Indeed, $y(t)$ is small for small values of t. The PLL reacts gradually to an abrupt change in $u(t)$, as it should. This limiting case holds.

3. The third limiting case is key for a correct PLL performance. If the PLL correctly measures a change in $u(t)$, it must eventually settle at the constant value $\tilde{u}$ as time progresses, albeit with some lag. Mathematically, this means that $y(t) \to \tilde{u}$ if $t \to \infty$. Let us see if this works for equation (8.2).

   If $t \to \infty$, the exponent there approaches zero, while trigonometric functions are limited to the -1 to 1 range. Therefore, the expression in the curly brackets approaches 1, and $y(t) \to u(t) = \tilde{u}$. This limiting case also holds.

Figure 8.3 shows a plot of $y(t)$ for $f_N = 15$ Hz, $\zeta = 0.2, \tilde{u} = 1$. It is consistent with the last two limiting cases: $y(t)$ approaches $\tilde{u}$ for large t; and it is small for small t, even though the input $u(t)$ jumps to $\tilde{u}$ for any positive value of t. The first limiting case ($y(t) = 0$ for $t < 0$) is also valid but lies outside of the plot domain for figure 8.3.

We have validated all three limiting cases for this problem. This increases our confidence in the result. It also supports the key design elements of the PLL algorithm: it is expected to work as desired.

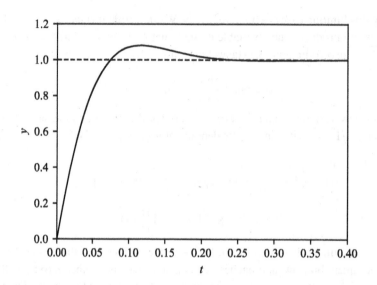

Figure 8.3
A PLL response to a step function input

8.4 Symmetry and Invariance

One trivial symmetry for equation (8.2) deals with flipping the sign of $u(t)$. The sign of $y(t)$ flips as a result. Another symmetry is not obvious here because equation (8.2) was derived for the jump in the step function that occurs precisely at $t = 0$. This becomes an implicit assumption in the solution, hiding the invariance with respect to a time shift. Mathematically, the time shift is expressed as follows:

$$
\begin{aligned}
y(t - t_0) = \left\{ 1 - \frac{e^{-2\pi\zeta f_N(t-t_0)}}{\sqrt{1 - \zeta^2}} \left[\sin\left(2\pi\zeta f_N \sqrt{1 - \zeta^2} \cdot (t - t_0) + \phi \right) \right.\right. \\
\left.\left. - 2\zeta \sin\left(2\pi\zeta f_N \sqrt{1 - \zeta^2} \cdot (t - t_0) \right) \right] \right\} u(t - t_0).
\end{aligned}
\tag{8.9}
$$

We are already familiar with the time-shift invariance from exercise 21 in chapter 3. The time-shift invariance is of great importance here: if we want to model a realistic motion profile as a sequence of small steps, such steps (with the possible exception of one of them) will not occur at $t = 0$ (see figure 8.2).

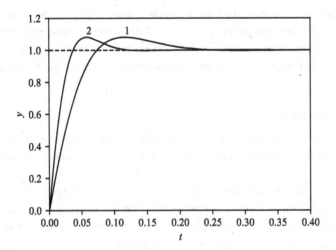

Figure 8.4
PLL responses for two different values of f_N

8.5 Scaling

Equation (8.2) exhibits two scaling behaviors. First, note that time t in the right-hand side always enters the equation in combination with the loop bandwidth f_N, except when used as the argument of the step function $u(t)$.

The step function is not affected by scaling t. The jump in $u(t)$ is instantaneous, and its shape does not change if we replace t with αt, as long as α is a positive constant.

Therefore, if we increase the loop bandwidth f_N, the plot for y as a function of t will be uniformly squeezed, and if we decrease the loop bandwidth, the plot will be stretched in the horizontal direction. Figure 8.4 illustrates this scaling behavior: it shows $y(t)$ for $\zeta = 0.2, \tilde{u} = 1$, and for two values of the loop bandwidth (curve 1: $f_{N1} = 15$ Hz; curve 2: $f_{N2} = 30$ Hz).

We have checked the limiting case for $t \rightarrow \infty$ above and concluded that after some time the measured value $y(t)$ will approach the true value $\tilde{u}$. Squeezing the plot of y in the horizontal direction will mean that this convergence to the true value will be achieved faster. Here scaling delivers a recommendation for one of the design parameters: increasing the loop bandwidth f_N speeds up the receiver reaction to a change in position.[3]

---

3. From this argument, it seems that increasing f_N is a no-brainer. In reality, it has a side effect: the GPS receiver becomes more susceptible to the noise in the signal. A designer must balance these two conflicting requirements to optimize performance.

The second scaling behavior stems from the linearity of the solution with respect to $u(t)$. Namely, scaling $u(t)$ by a constant multiplier scales the solution $y(t)$ accordingly. This, along with the time-shift invariance, is key for our ability to model any motion profile as a sequence of variable-size small steps and for using equation (8.2) to predict the resulting measurements.

8.6   Order of Magnitude Estimates

In the previous section, we concluded that the loop bandwidth f_N is the key parameter that determines how fast the PLL reacts to an external change. Here we make a rough estimate for the reaction time.

Upon examining equation (8.2), we see that the exponent drives the convergence of $y(t)$ to $u(t)$ (see also limiting case 3 in section 8.3). The estimate for the convergence time τ is given by equating the argument of the exponent to -1; over that time the value of the exponent decreases by a factor of $e \approx 2.7$, which is a significant change:

$$\tau = \frac{1}{2\pi \zeta f_N}. \tag{8.10}$$

Even though we obtained this estimate assuming a step-function input, it should be valid for other inputs as well. Indeed, we have speculated that the time-shift invariance and the scaling allow us to model any input as a sequence of small steps. If the convergence for each such small step takes time τ, then $y(t)$ will roughly lag the input motion profile by this time, regardless of the shape of the input.

8.7   Successive Approximations

In the last section, we obtained a rough estimate for the time of convergence τ. This is a good start, but for designing a receiver one needs to do better than that. The manufacturer has to provide data to customers to guarantee a required performance level. Accurate predictions are also important for testing a real device. In this section, we answer the following question: how long does it take for the PLL to get within a small vicinity of $\tilde{u}$ after a jump in $u(t)$?

From figure 8.3, you can see that $y(t)$ oscillates near the correct value $\tilde{u}$. For this problem, we solve for the time moment when $y(t)$ first approaches $\tilde{u}$. Mathematically, this means that we need to solve the following equation for t:

$$y(t) = (1 - \delta)u(t), \tag{8.11}$$

while keeping in mind that we seek the first solution (one that corresponds to the smallest value of t). For example, if $\delta = 0.05$, the solution of this equation will determine the time that is required to get within 5 percent of the true change in the distance between the satellite and the GPS receiver, and so on.

Substituting $y(t)$ from equation (8.2) yields

$$\frac{e^{-2\pi\zeta f_N t}}{\sqrt{1 - \zeta^2}} \left(\sin\left(2\pi\zeta f_N \sqrt{1 - \zeta^2} \cdot t + \phi\right) - 2\zeta \sin\left(2\pi\zeta f_N \sqrt{1 - \zeta^2} \cdot t\right) \right) = \delta. \qquad (8.12)$$

This is a transcendental and quite complex equation, and there is no closed-form solution for it. We turn to the method of successive approximations (MSA) to solve it. To apply the MSA, we transform the last equation as follows:

$$\sin\left(2\pi\zeta f_N \sqrt{1 - \zeta^2} \cdot t + \phi\right) - 2\zeta \sin\left(2\pi\zeta f_N \sqrt{1 - \zeta^2} \cdot t\right) = \delta \sqrt{1 - \zeta^2} e^{2\pi\zeta f_N t}. \qquad (8.13)$$

Next, we introduce new notation to simplify the last equation:

$$\begin{aligned} p &= 1, \\ q &= -2\zeta, \\ x &= 2\pi\zeta f_N \sqrt{1 - \zeta^2} \cdot t, \\ c &= \delta \sqrt{1 - \zeta^2} e^{2\pi\zeta f_N t}. \end{aligned} \qquad (8.14)$$

In these notations, equation (8.13) looks exactly like equation (A.62) solved in section A.14:

$$p\sin(x + \phi) + q\sin x = c. \qquad (8.15)$$

Its solution is as follows:

$$x = -\tan^{-1}\frac{p\sin\phi}{p\cos\phi + q} + (-1)^n \sin^{-1}\left(\frac{c}{\sqrt{p^2 + 2pq\cos\phi + q^2}}\right) + n\pi. \qquad (8.16)$$

We have already established that $\sin\phi = \sqrt{1 - \zeta^2}, \cos\phi = \zeta$. We use these values and definitions of p, q to simplify various expressions in the solution. We substitute p, q from equations (8.14) in the inverse tangent in equation (8.16) and then use $\zeta = \cos\phi$:

$$\begin{aligned} \tan^{-1}\frac{p\sin\phi}{p\cos\phi + q} &= \tan^{-1}\frac{1 \cdot \sin\phi}{\cos\phi - 2\zeta} \\ &= \tan^{-1}\frac{\sin\phi}{\cos\phi - 2\cos\phi} \\ &= -\tan^{-1}\frac{\sin\phi}{\cos\phi} \\ &= -\phi. \end{aligned} \qquad (8.17)$$

Next we deal with the radical in equation (8.16) and do the same:

Table 8.1
Time required to reach the vicinity of the exact value by the PLL

Iteration	S	$f_N t$
0	0.0	0.25
1	0.021473	0.271473
2	0.023623	0.273623
3	0.02385	0.27385
4	0.023874	0.273874
5	0.023877	0.273877

$$\sqrt{p^2 + 2pq\cos\phi + q^2} = \sqrt{1 + 2(-2\zeta)\cos\phi + (-2\zeta)^2}$$
$$= \sqrt{1 + 2(-2\cos\phi)\cos\phi + (-2\cos\phi)^2} \qquad (8.18)$$
$$= 1.$$

Now equation (8.16) is substantially simplified:

$$x = \phi + (-1)^n \sin^{-1}(c) + n\pi. \qquad (8.19)$$

From now on we use $n = 0$ for the first approach to the stationary state. We recall the definitions for x and c to get

$$2\pi\zeta f_N \sqrt{1 - \zeta^2} \cdot t = \phi + \sin^{-1}\left(\delta\sqrt{1 - \zeta^2}e^{2\pi\zeta f_N t}\right). \qquad (8.20)$$

Now we can solve for $f_N t$ in the left-hand side:[4]

$$f_N t = \frac{\phi}{2\pi\zeta\sqrt{1 - \zeta^2}} + \frac{1}{2\pi\zeta\sqrt{1 - \zeta^2}} \sin^{-1}\left(\delta\sqrt{1 - \zeta^2}e^{2\pi\zeta f_N t}\right). \qquad (8.21)$$

This form of the equation for $f_N t$ is suitable for applying the MSA. Indeed, even though the right-hand side contains $f_N t$, it also has a small parameter[5] δ.

Let us assume $\zeta = 1/\sqrt{2}$, which corresponds to $\phi = \pi/4$. We also use $\delta = 0.01$. Results from the first five iterations are presented in table 8.1, where we denoted for brevity

---

4. We solve for the dimensionless combination $f_N t$ and not for t proper. Because of the scaling, any value of $f_N t$ corresponds to a family of solutions for different pairs of f_N and t, and any results depend only on their product $f_N t$.

5. A word of caution: the small parameter here is multiplied by the exponent, and we must keep an eye on the solution to make sure that a large value of the exponent does not invalidate the effect of the small parameter.

$$S = \sin^{-1}\left(\delta\sqrt{1-\zeta^2}\,e^{2\pi\zeta f_N t}\right).$$ (8.22)

We see that the MSA delivers a solution for this rather complex problem.

8.8 Summary

The previous section completes our investigation of the GPS receiver tracking problem. We applied a suite of analyses to a solution that is the product of a complex and lengthy derivation. We have validated several key features of this solution. We made sure that our tracking algorithm will faithfully follow any variations in the distance between the GPS satellite and the receiver. We also got some helpful cues for designing a receiver if we want it to react faster to any changes in the signal. We laid out a foundation to extend a solution for the step-function input to an input of any shape, as enabled by scaling and the time-shift invariance. Finally, we solved for the time the PLL takes to approximate the input value.

The tools we have learned in chapters 1 through 6 have paved a path from an unwieldy mathematical equation to a better understanding of an engineering problem. It is the goal of this book to teach you how to chart such a path for any problem in hand. The success of scientists and engineers depends in part on their ability to make sense of complex phenomena that are often described by baffling mathematical equations. Fortunately, every formula has a meaning, and deep down it tells a story about the things it describes. This book arms you with the tools to pry open the shell of formal mathematics and to illuminate the concepts that lie within. Often this is the only way to move forward in your work. An intuitive, commonsense understanding of a problem is also a great help in communicating your results and ideas to others.

The tools presented in this book go beyond using canned mathematical recipes. They are a great help in a deep and unrestrained exploration of your field of work. Whether designing a new wireless telecommunications system, developing an artificial intelligence application, or reviewing data from a clinical trial, you will be able to get to a better result faster and easier.

A Problems and Solutions

This appendix contains various problems that are used over and over in chapters 1 through 6 as fodder for exercises or to illustrate the material. Each problem is the subject of a separate section below. Full solutions, if present, are for your reference; chapters 1 through 6 often refer to these problems and the final results but rarely dive into the derivations.

A.1 Two Hikers on a Trail

Problem

Consider two hikers starting to walk toward each other from the opposite ends of a trail (figure A.1). One hiker maintains speed V_1, and the other maintains speed V_2. The total length of the trail is D. What distance will each of them cover before meeting on the trail?

Solution

We denote the respective positions of both hikers on the trail as x_1 and x_2. We also assume that the hikers start walking at $t = 0$. The hikers' positions as functions of time are given by

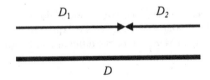

Figure A.1
Distances traveled by two hikers on a trail

$$x_1 = V_1 t,$$
$$x_2 = D - V_2 t. \tag{A.1}$$

Note that the second hiker is at $x_2 = D$ for $t = 0$. The minus sign in the second equation means that the second hiker is moving toward the first one.

The two hikers meet when $x_1 = x_2$. From equations (A.1) we get

$$V_1 T = D - V_2 T, \tag{A.2}$$

where T denotes the time for the hikers to meet. This yields

$$T = \frac{D}{V_1 + V_2}. \tag{A.3}$$

During this time, the first hiker will walk the distance

$$D_1 = V_1 T = \frac{D V_1}{V_1 + V_2}. \tag{A.4}$$

Analogously, the second hiker will walk the distance

$$D_2 = V_2 T = \frac{D V_2}{V_1 + V_2}. \tag{A.5}$$

A.2 A Riverboat

Problem

A riverboat travels from town A to town B in T_{AB} hours and from town B to town A in T_{BA} hours. Town B is located upstream, town A is downstream, and the river's current is the cause for the difference in the travel times. How much time would it take a raft to go from town B to town A?

Solution

We denote the desired travel time on a raft as T_r. We also denote the speed of the river current as V_r (fittingly, the subscript is also a notation for the speed of a raft), the speed of the boat with respect to the water as V_b, and the distance between the two towns as D.

In one case, the speed of the boat with respect to the banks of the river will be slowed down by the river's current, and in the other case it will be sped up by it. Mathematically, this is expressed by the fact that the speed of the boat with respect to the banks of the river will be the difference or the sum of variables V_b and V_r:

$$V_b - V_r = \frac{D}{T_{AB}},$$
$$V_b + V_r = \frac{D}{T_{BA}}. \tag{A.6}$$

We subtract these two equations to compute the speed of the river's current (net of the boat's speed):

$$V_r = \frac{1}{2}\left(\frac{D}{T_{BA}} - \frac{D}{T_{AB}}\right) = \frac{D(T_{AB} - T_{BA})}{2 T_{AB} T_{BA}}. \tag{A.7}$$

The time to cover distance D with this speed is computed as

$$T_r = \frac{D}{V_r} = \frac{2T_{AB}T_{BA}}{T_{AB} - T_{BA}}. \tag{A.8}$$

A.3 The Intersections between a Circle and a Straight Line

Problem

We are given the equations for a circle ($x^2 + y^2 = R^2$) and for a straight line ($y = px + q$). We need to find the coordinates of the intersections of these two curves (if any). Figure A.2 illustrates this problem.

Solution

At the intersections, x and y satisfy both equations:

$$x^2 + y^2 = R^2,$$
$$y = px + q. \tag{A.9}$$

First, let's think about what happens for different values of parameters R, p, and q. Parameter R defines the circle radius, and q defines the point of intersection between the straight line and the vertical axis. Parameter p defines the slope of the straight line. If we vary q and fix the values of R and p, the straight line would shift up or down with respect to the circle. In figure A.2, there are two intersections between the straight line and the circle. If the line moves up, the spacing between the two intersections will become smaller; at some value of q, they will merge; for larger values of q, the line and the circle will no longer intersect. This would mean that equations (A.9) have no roots.

This behavior is similar to what happens with quadratic equations. Indeed, a quadratic equation with a positive discriminant has two real roots. As we vary its coefficients in such a way that the discriminant decreases, the two roots move closer to each other. At some point (when the discriminant is zero) the two roots have the same value. If we vary the coefficients further and the discriminant

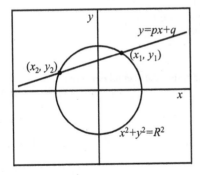

Figure A.2

The intersections between a circle and a straight line

becomes negative, the equation no longer has two real roots. (Of course, it still has two complex roots.)

The similarity between the behavior of roots for this problem and the behavior of roots of quadratic equations is not accidental. Indeed, let us solve equations (A.9) for x. We substitute the value of y from the second equation in the first one to get

$$x^2 + (px + q)^2 = R^2. \tag{A.10}$$

After expanding the square and collecting the terms, we do get a quadratic equation for x:

$$(1 + p^2)x^2 + 2pqx + q^2 - R^2 = 0. \tag{A.11}$$

The roots are given by

$$x_{1,2} = \frac{-pq \pm \sqrt{p^2q^2 - (q^2 - R^2)(1 + p^2)}}{1 + p^2}, \tag{A.12}$$

where subscripts $1, 2$ correspond to the $\pm$ signs in the right-hand side. The expression under the square root can be simplified to get

$$x_{1,2} = \frac{-pq \pm \sqrt{(1 + p^2)R^2 - q^2}}{1 + p^2}. \tag{A.13}$$

These roots exist if the discriminant is nonnegative: $(1 + p^2)R^2 - q^2 \geq 0$.

A.4 The Intersections between a Circle and an Ellipse

Problem

The following equations define a circle and an ellipse that are both centered at the origin (see figure A.3):

$$x^2 + y^2 = R^2,$$
$$\frac{x^2}{R_x^2} + \frac{y^2}{R_y^2} = 1. \tag{A.14}$$

Find the coordinates of the intersections of these two curves (if any).

Solution

We express x^2 from the first equation of (A.14) and plug it into the second equation:

$$x^2 = R^2 - y^2,$$
$$\frac{R^2 - y^2}{R_x^2} + \frac{y^2}{R_y^2} = 1. \tag{A.15}$$

Now we can solve the last equation for y^2:

$$R_y^2(R^2 - y^2) + R_x^2 y^2 = R_x^2 R_y^2,$$
$$y^2 = R_y^2 \frac{R_x^2 - R^2}{R_x^2 - R_y^2}. \tag{A.16}$$

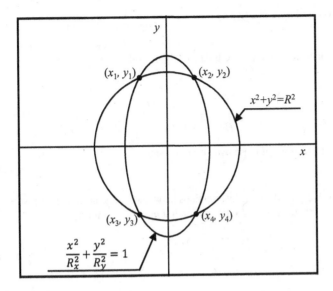

Figure A.3
The intersections between a circle and an ellipse

Next, we plug this result into the first equation in (A.15) to solve for x^2:

$$
\begin{aligned}
x^2 &= R^2 - y^2 \\
&= R^2 - R_y^2 \frac{R_x^2 - R^2}{R_x^2 - R_y^2} \\
&= \frac{R^2 R_x^2 - R^2 R_y^2 - R_y^2 R_x^2 + R^2 R_y^2}{R_x^2 - R_y^2} \\
&= R_x^2 \frac{R_y^2 - R^2}{R_y^2 - R_x^2}.
\end{aligned}
\tag{A.17}
$$

The solutions for x and y are given by

$$
\begin{aligned}
x &= \pm \sqrt{R_x^2 \frac{R_y^2 - R^2}{R_y^2 - R_x^2}}, \\
y &= \pm \sqrt{R_y^2 \frac{R_x^2 - R^2}{R_x^2 - R_y^2}}.
\end{aligned}
\tag{A.18}
$$

If both expressions under the square root are nonnegative, there are four possible combinations of the plus and minus signs, corresponding to four intersections between the circle and the ellipse.

A.5 The Intersections between a Circle and a Hyperbola

Problem

The following equations define a circle and a hyperbola (see figure A.4):

$$x^2 + y^2 = R^2,$$
$$\frac{x^2}{H_x^2} - \frac{y^2}{H_y^2} = 1. \tag{A.19}$$

Find the coordinates of the intersections of these two curves (if any).

Solution

The solution is similar to how we solved for the intersections between a circle and an ellipse (section A.4). We express x^2 from the first equation of (A.19) and plug it into the second equation:

$$x^2 = R^2 - y^2,$$
$$\frac{R^2 - y^2}{H_x^2} - \frac{y^2}{H_y^2} = 1. \tag{A.20}$$

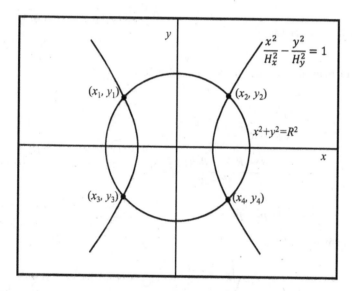

Figure A.4
The intersections between a circle and a hyperbola

Next, we solve the last equation for y^2:

$$y^2 = H_y^2 \frac{R^2 - H_x^2}{H_x^2 + H_y^2}. \tag{A.21}$$

Now we plug this result into the first equation of (A.20) to solve for x^2:

$$
\begin{aligned}
x^2 &= R^2 - y^2 \\
&= R^2 - H_y^2 \frac{R^2 - H_x^2}{H_x^2 + H_y^2} \\
&= \frac{R^2 H_x^2 + R^2 H_y^2 - R^2 H_y^2 + H_x^2 H_y^2}{H_x^2 + H_y^2} \\
&= H_x^2 \frac{R^2 + H_y^2}{H_x^2 + H_y^2}.
\end{aligned}
\tag{A.22}
$$

This yields the following solutions for x and y:

$$
\begin{aligned}
x &= \pm \sqrt{H_x^2 \frac{R^2 + H_y^2}{H_x^2 + H_y^2}}, \\
y &= \pm \sqrt{H_y^2 \frac{R^2 - H_x^2}{H_x^2 + H_y^2}}.
\end{aligned}
\tag{A.23}
$$

If the expressions under the square roots are nonnegative, there are four possible combinations of plus and minus signs, corresponding to four intersections between the circle and the hyperbola.

A.6 The Intersections between a Circle and a Parabola

Problem

The following equations define a circle and a parabola (see figure A.5):

$$
\begin{aligned}
x^2 + y^2 &= R^2, \\
y &= g x^2 + y_0.
\end{aligned}
\tag{A.24}
$$

Find the coordinates of the intersections of these two curves (if any).

Solution

We substitute x^2 from the first equation in the second one to get a quadratic equation for y:

$$y = g(R^2 - y^2) + y_0. \tag{A.25}$$

We rearrange the terms to get

$$g y^2 + y - g R^2 - y_0 = 0. \tag{A.26}$$

For a nonnegative discriminant, the solutions are given by the quadratic formula:

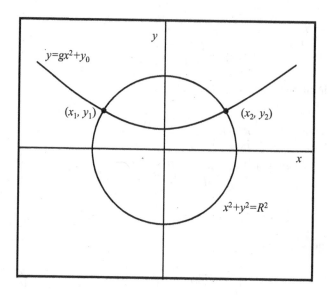

Figure A.5
The intersections between a circle and a parabola

$$y_{1,2} = \frac{-1 \pm \sqrt{1 + 4g\left(gR^2 + y_0\right)}}{2g}, \tag{A.27}$$

where subscripts $1, 2$ correspond to the $\pm$ signs in the right-hand side. We substitute these solutions in the first equation of (A.24) to get

$$x_{1,2} = +\sqrt{R^2 - y_{1,2}^2},$$
$$x_{3,4} = -\sqrt{R^2 - y_{1,2}^2}. \tag{A.28}$$

In addition to the previous requirement of having a nonnegative discriminant, we note that the solutions are valid only for such values of y that satisfy condition $R^2 \geq y_{1,2}^2$ or, equivalently, $R \geq |y_{1,2}|$. Therefore, if the discriminant $1 + 4g(gR^2 + y_0)$ is negative, equations (A.24) have no solutions; if the discriminant is nonnegative and only one of the values of $y_{1,2}$ satisfies condition $R \geq |y_{1,2}|$, equations (A.24) have two solutions; finally, if both values of $y_{1,2}$ satisfy condition $R \geq |y_{1,2}|$, equations (A.24) have four solutions.

A.7 The Product of Two Linear Expressions

Problem

Solve for x:

$$(x - a)(x - b) = d. \tag{A.29}$$

Solution

We expand the expression in the brackets and regroup the terms to get

$$x^2 - x(a+b) + ab - d = 0. \tag{A.30}$$

The discriminant of this quadratic equation can be simplified by expanding the square and regrouping the terms:

$$(a+b)^2 - 4(ab - d) = (a-b)^2 + 4d. \tag{A.31}$$

With this expression for the discriminant, the solution of equation (A.29) is as follows:

$$x_{1,2} = \frac{(a+b) \pm \sqrt{(a-b)^2 + 4d}}{2}, \tag{A.32}$$

where subscripts 1, 2 correspond to the ± signs in the right-hand side.

A.8   The Sum of an Unknown and Its Reciprocal

Problem

Solve for x:

$$x + \frac{1}{x} = d. \tag{A.33}$$

Solution

We multiply both sides of equation (A.33) by x to get

$$x^2 - dx + 1 = 0. \tag{A.34}$$

This is a quadratic equation, and its solutions are

$$x_{1,2} = \frac{d \pm \sqrt{d^2 - 4}}{2}, \tag{A.35}$$

where subscripts 1, 2 correspond to the ± signs in the right-hand side.

A.9   The Difference of an Unknown and Its Reciprocal

Problem

Solve for x:

$$x - \frac{1}{x} = d. \tag{A.36}$$

Solution

We multiply both sides of equation (A.36) by x to get

$$x^2 - dx - 1 = 0. \tag{A.37}$$

This is a quadratic equation, and its solutions are

$$x_{1,2} = \frac{d \pm \sqrt{d^2 + 4}}{2}, \tag{A.38}$$

where subscripts 1, 2 correspond to the $\pm$ signs in the right-hand side.

A.10   The Sum of Two Ratios

Problem

Solve for x:

$$\frac{1}{x-a} + \frac{1}{x-b} = d. \tag{A.39}$$

Solution

We multiply both sides of equation (A.39) by $(x-a)(x-b)$ to get

$$(x-b) + (x-a) = d(x-a)(x-b). \tag{A.40}$$

After expanding the expression in the brackets and regrouping the terms, we obtain a quadratic equation:

$$dx^2 - x(2 + d(a+b)) + (a+b) + dab = 0. \tag{A.41}$$

The discriminant of this equation can be simplified by expanding the square and regrouping the terms:

$$\begin{aligned}
(2 + d(a+b))^2 - 4d\,((a+b) + dab) = \\
4 + d^2(a+b)^2 + 4d(a+b) - 4d\,((a+b) + dab) = \\
d^2(a+b)^2 - 4d^2ab + 4 = \\
d^2(a-b)^2 + 4.
\end{aligned} \tag{A.42}$$

With this expression for the discriminant, the solution of equation (A.39) is as follows:

$$x_{1,2} = \frac{2 + d(a+b) \pm \sqrt{d^2(a-b)^2 + 4}}{2d}, \tag{A.43}$$

where subscripts 1, 2 correspond to the $\pm$ signs in the right-hand side.

A.11   The Sum of Two Scaled Ratios

Problem

Solve for x:

$$\frac{p}{x-a} + \frac{q}{x-b} = d. \tag{A.44}$$

Solution

The solution of this equation is analogous to that for equation (A.39). We multiply both sides of equation (A.44) by $(x-a)(x-b)$ to get

$$p(x-b) + q(x-a) = d(x-a)(x-b). \tag{A.45}$$

After expanding the expression in the brackets and regrouping the terms, we obtain a quadratic equation:

$$dx^2 - x((p+q) + d(a+b)) + (qa + pb) + dab = 0. \tag{A.46}$$

The discriminant of this equation can be simplified by expanding the square and regrouping the terms:

$$
\begin{aligned}
((p+q) + d(a+b))^2 &- 4d\,((qa+pb) + dab) = \\
d^2(a+b)^2 + 2d(p+q)(a+b) + (p+q)^2 &- 4d(pb+qa) - 4d^2 ab = \\
d^2(a-b)^2 + (p+q)^2 &+ 2d(pa + qb - pb - qa) = \\
d^2(a-b)^2 + (p+q)^2 &+ 2d(a-b)(p-q).
\end{aligned}
\tag{A.47}
$$

With this expression for the discriminant, the solution of equation (A.44) is as follows:

$$x_{1,2} = \frac{(p+q) + d(a+b) \pm \sqrt{d^2(a-b)^2 + (p+q)^2 + 2d(a-b)(p-q)}}{2d}, \tag{A.48}$$

where subscripts 1, 2 correspond to the ± signs in the right-hand side.

A.12 The Difference of Two Ratios

Problem

Solve for x:

$$\frac{1}{x-a} - \frac{1}{x-b} = d. \tag{A.49}$$

Solution

The solution of this equation is analogous to that for equation (A.39). We multiply both sides of equation (A.49) by $(x-a)(x-b)$ to get

$$a - b = d(x-a)(x-b). \tag{A.50}$$

After collecting the terms, we arrive at a quadratic equation:

$$dx^2 - dx(a+b) + dab - (a-b) = 0. \tag{A.51}$$

Its solution is given by the quadratic formula:

$$x_{1,2} = \frac{d(a+b) \pm \sqrt{d^2(a+b)^2 - 4d\,(dab - (a-b))}}{2d}, \tag{A.52}$$

where subscripts 1, 2 correspond to the ± signs in the right-hand side. The discriminant can be simplified to get the final result:

$$x_{1,2} = \frac{d(a+b) \pm \sqrt{d^2(a-b)^2 + 4d(a-b)}}{2d}. \tag{A.53}$$

A.13   The Sum of Trigonometric Functions, 1st Version

Problem

Solve for x:

$$a \sin x + b \cos x = c. \tag{A.54}$$

Solution

We divide this equation by $\sqrt{a^2 + b^2}$ to get

$$g \sin x + h \cos x = \frac{c}{\sqrt{a^2 + b^2}}, \tag{A.55}$$

where we denoted

$$g = \frac{a}{\sqrt{a^2 + b^2}},$$
$$h = \frac{b}{\sqrt{a^2 + b^2}}. \tag{A.56}$$

Note that $g^2 + h^2 = 1$. This means that we can always select such a variable α that $g = \cos \alpha$ and $h = \sin \alpha$. Specifically, we set[1]

$$\alpha = \tan^{-1} \frac{h}{g} = \tan^{-1} \frac{b}{a}. \tag{A.57}$$

Substituting of expressions $g = \cos \alpha$ and $h = \sin \alpha$ in equation (A.55) yields

$$\cos \alpha \sin x + \sin \alpha \cos x = \frac{c}{\sqrt{a^2 + b^2}}. \tag{A.58}$$

Using the trigonometric identity[2] for the sine of a sum, we get

$$\sin(x + \alpha) = \frac{c}{\sqrt{a^2 + b^2}}. \tag{A.59}$$

The solution of this equation is as follows:

$$x = -\alpha + (-1)^n \sin^{-1}\left(\frac{c}{\sqrt{a^2 + b^2}}\right) + n\pi, \tag{A.60}$$

where n is an integer. We recall the expression for α to get the final result:

$$x = -\tan^{-1} \frac{b}{a} + (-1)^n \sin^{-1}\left(\frac{c}{\sqrt{a^2 + b^2}}\right) + n\pi. \tag{A.61}$$

1. Depending on the signs of a and b, we may have to use $\alpha = \tan^{-1}(b/a) + \pi$ to get the correct result.
2. The sine of a sum of two angles: $\sin \gamma \cos \beta + \sin \beta \cos \gamma = \sin(\gamma + \beta)$.

A.14 The Sum of Trigonometric Functions, 2nd Version

Problem

Solve for x:

$$p \sin(x + \phi) + q \sin x = c. \tag{A.62}$$

Solution

We use the trigonometric identity for the sine of a sum (see footnote 2 in section A.13) to get

$$p \sin x \cos \phi + p \cos x \sin \phi + q \sin x = c. \tag{A.63}$$

After grouping the terms, this equation is as follows:

$$(p \cos \phi + q) \sin x + p \sin \phi \cos x = c. \tag{A.64}$$

We introduce new notations:

$$a = p \cos \phi + q,$$
$$b = p \sin \phi. \tag{A.65}$$

In these notations, equation (A.64) looks exactly the same as the one solved in section A.13. Leveraging that result, we obtain a solution[3] through a, b, and c:

$$x = -\tan^{-1} \frac{b}{a} + (-1)^n \sin^{-1} \left(\frac{c}{\sqrt{a^2 + b^2}} \right) + n\pi. \tag{A.66}$$

Finally, we recall the definitions for a and b to express this solution through the original parameters p, q, and c:

$$x = -\tan^{-1} \frac{p \sin \phi}{p \cos \phi + q} + (-1)^n \sin^{-1} \left(\frac{c}{\sqrt{p^2 + 2pq \cos \phi + q^2}} \right) + n\pi. \tag{A.67}$$

A.15 The Ratio of Cosines

Problem

Solve for x:

$$\frac{\cos(\alpha + x)}{\cos(\alpha - x)} = \frac{p}{q}. \tag{A.68}$$

---

3. Depending on the signs of a and b, we may have to use $\tan^{-1}(b/a) + \pi$ instead of $\tan^{-1}(b/a)$ to get the correct result.

Solution

We use the identities for the cosine of a sum or of a difference of two angles[4] to expand the left-hand side:

$$\frac{\cos\alpha\cos x - \sin\alpha\sin x}{\cos\alpha\cos x + \sin\alpha\sin x} = \frac{p}{q}. \tag{A.69}$$

Next, we divide both the numerator and denominator of the left-hand side by $\cos\alpha \cdot \cos x$ (assuming that both are nonzero; the case of $\cos\alpha \cdot \cos x = 0$ can be handled separately):

$$\frac{1 - \tan\alpha\tan x}{1 + \tan\alpha\tan x} = \frac{p}{q}. \tag{A.70}$$

We multiply by the denominators and solve for $\tan x$ to get

$$\tan x = \cot\alpha \cdot \frac{q-p}{q+p}. \tag{A.71}$$

Therefore,

$$x = \tan^{-1}\left(\cot\alpha \cdot \frac{q-p}{q+p}\right) + n\pi, \tag{A.72}$$

where n is an integer.

A.16   Blending Two Syrups

Problem

A portion of a syrup is characterized by its total mass and by the concentration of sugar. Concentration p means that there are p kilograms of sugar per unit mass (one kilogram) of syrup. In this problem, we start with two syrups. The first syrup has mass m_1 and sugar concentration p_1. The second syrup has mass m_2 and sugar concentration p_2. What is the concentration of sugar in the mix of these two syrups?

Solution

The solution is based on the conservation of mass, both for the syrup and for the sugar in it. The total mass m_{12} is the sum of the masses of the two original syrups:

$$m_{12} = m_1 + m_2. \tag{A.73}$$

The amount of sugar in each of the original syrups and in their mix is the product of the concentration and the total mass of that syrup. Since the amount of sugar is conserved, we get another equation:

$$p_{12}m_{12} = p_1 m_1 + p_2 m_2, \tag{A.74}$$

where p_{12} is the concentration of sugar in the mix.

The last two equations must be solved for p_{12}. We substitute m_{12} from the first equation in the second one. This yields the desired formula for the sugar concentration in the mix of the two syrups:

---

4. The cosine of a sum or difference of two angles: $\cos(\gamma \pm \beta) = \cos\gamma\cos\beta \mp \sin\gamma\sin\beta$.

$$p_{12} = \frac{p_1 m_1 + p_2 m_2}{m_1 + m_2}. \tag{A.75}$$

A.17 Blending Three Syrups

Problem

In this problem, we mix three syrups that have masses m_1, m_2, and m_3 and concentrations p_1, p_2, and p_3. What is the concentration of sugar in the mix of these three syrups?

Solution

The solution is completely analogous to the case of two syrups in section A.16 above. We start from

$$m_{123} = m_1 + m_2 + m_3,$$
$$p_{123} m_{123} = p_1 m_1 + p_2 m_2 + p_3 m_3. \tag{A.76}$$

This yields

$$p_{123} = \frac{p_1 m_1 + p_2 m_2 + p_3 m_3}{m_1 + m_2 + m_3}. \tag{A.77}$$

A.18 Draining a Pool Using Two Pumps

Problem

A pool full of water must be drained for winter using two pumps. The first pump can drain this pool in T_1 hours, and the second pump can do it in T_2 hours. How long would it take to drain the pool if we use both pumps?

Solution

We denote the total amount of water in the pool as M, measured in cubic meters. The first pump can drain $R_1 = M/T_1$ cubic meters per hour, and the second pump can drain $R_2 = M/T_2$ cubic meters per hour. If both pumps work together, they can drain $R_1 + R_2$ cubic meters per hour. At this rate, it will take

$$T_{12} = \frac{M}{R_1 + R_2} \tag{A.78}$$

hours to drain all the water in the pool. Substituting the values for R_1 and R_2, we get

$$T_{12} = \frac{M}{\frac{M}{T_1} + \frac{M}{T_2}} = \frac{T_1 T_2}{T_1 + T_2}. \tag{A.79}$$

A.19 Draining a Pool Using Three Pumps

Problem

A pool full of water must be drained using three pumps. Individually, these pumps can drain this pool in T_1, T_2, and T_3 hours. How long would it take to drain the pool if we use all three pumps?

Solution

The solution is analogous to that for the problem in the previous section. The result is as follows:

$$T_{123} = \frac{T_1 T_2 T_3}{T_1 T_2 + T_1 T_3 + T_2 T_3}. \tag{A.80}$$

A.20 The Sum of Two Radicals

Problem

Solve for x, assuming $a > 0, b > 0$:

$$\sqrt{a+x} + \sqrt{a-x} = b. \tag{A.81}$$

Solution

Both sides of this equation are positive. We square this equation to get

$$2a + 2\sqrt{a^2 - x^2} = b^2. \tag{A.82}$$

Next, we solve for the square root in the left-hand side and square the resulting equation again:

$$\sqrt{a^2 - x^2} = \frac{b^2}{2} - a,$$
$$a^2 - x^2 = \left(\frac{b^2}{2} - a\right)^2. \tag{A.83}$$

This yields

$$x^2 = a^2 - \left(\frac{b^2}{2} - a\right)^2. \tag{A.84}$$

Finally, we compute the square root of the last equation to get the solutions for x:

$$x = \pm\sqrt{a^2 - \left(\frac{b^2}{2} - a\right)^2}. \tag{A.85}$$

A.21 The Difference of Two Radicals

Problem

Solve for x, assuming $a > 0$:

$$\sqrt{a+x} - \sqrt{a-x} = b. \tag{A.86}$$

Compared with the previous problem, we drop the requirement that $b > 0$ here and allow b to be zero or negative.

Solution

First, we solve it for nonnegative values of x; the case of $x < 0$ will be considered separately. If $x \geq 0$, the left-hand side of this equation is nonnegative because the first term in the left-hand side $\sqrt{a+x}$ is greater than or equal to the second term $\sqrt{a-x}$. In this case, b must be nonnegative as well, and the solution is analogous to the previous problem. We square equation (A.86) to get

$$2a - 2\sqrt{a^2 - x^2} = b^2. \tag{A.87}$$

Next, we solve for the square root and square the resulting equation again:

$$\sqrt{a^2 - x^2} = -\frac{b^2}{2} + a,$$
$$a^2 - x^2 = \left(\frac{b^2}{2} - a\right)^2. \tag{A.88}$$

This yields

$$x^2 = a^2 - \left(\frac{b^2}{2} - a\right)^2. \tag{A.89}$$

Finally, we compute the square root of the last equation to get the solution for x:

$$x = \sqrt{a^2 - \left(\frac{b^2}{2} - a\right)^2}. \tag{A.90}$$

Here we selected only the plus sign for the square root because we have assumed that $x \geq 0$.

For a negative x, the situation is flipped: the left-hand side of this equation is negative because the first term in the left-hand side $\sqrt{a + x}$ is *smaller* than the second term $\sqrt{a - x}$. In this case, b must be negative as well. The solution follows the same path and arrives at the same expression for x^2:

$$x^2 = a^2 - \left(\frac{b^2}{2} - a\right)^2. \tag{A.91}$$

We compute the square root of the last equation to get two possible solutions for x, where we now select the *minus* sign (recall that this is the case where $x < 0$):

$$x = -\sqrt{a^2 - \left(\frac{b^2}{2} - a\right)^2}. \tag{A.92}$$

Note that solutions (A.90) and (A.92) apply respectively to cases $b \geq 0$ and $b < 0$.

Oddly, the functional form of solutions for x for this problem is exactly the same as the functional form of solution (A.85) of the problem in section (A.20), even though the starting equations are different. How can that be? The answer is given in section 2.8, which explores limiting cases for these two problems.

A.22 The Sum of Two Rational Functions

Problem

Solve for x:

$$\frac{x - a}{x - b} + \frac{x - b}{x - a} = c. \tag{A.93}$$

Solution

We multiply this equation by both denominators:

$$(x - a)^2 + (x - b)^2 = c(x - a)(x - b). \tag{A.94}$$

After expanding the squares and the product and collecting the terms, we arrive at a quadratic equation for x:

$$(2 - c)x^2 - (2 - c)(a + b)x + \left(a^2 + b^2 - cab\right) = 0. \tag{A.95}$$

Its solution is given by the quadratic formula:

$$x_{1,2} = \frac{(2 - c)(a + b) \pm \sqrt{(2 - c)^2 (a + b)^2 - 4(2 - c)(a^2 + b^2 - cab)}}{2(2 - c)}, \tag{A.96}$$

where subscripts 1, 2 correspond to the $\pm$ signs in the right-hand side.

After some algebra, the discriminant can be simplified to

$$(2 - c)^2 (a + b)^2 - 4(2 - c)(a^2 + b^2 - cab) = (c^2 - 4)(a - b)^2. \tag{A.97}$$

This yields the final result:

$$x_{1,2} = \frac{(2 - c)(a + b) \pm (a - b)\sqrt{c^2 - 4}}{2(2 - c)}. \tag{A.98}$$

A.23 The Difference of Two Rational Functions

Problem

Solve for x:

$$\frac{x - a}{x - b} - \frac{x - b}{x - a} = c. \tag{A.99}$$

Solution

The solution of this problem is analogous to that of the previous problem. We multiply this equation by both denominators:

$$(x - a)^2 - (x - b)^2 = c(x - a)(x - b). \tag{A.100}$$

After expanding the squares and the product and collecting the terms, we arrive at a quadratic equation for x:

$$cx^2 + (-c(a + b) + 2(a - b))x + \left(cab - a^2 + b^2\right) = 0. \tag{A.101}$$

Its solution is given by the quadratic formula:

$$x_{1,2} = \frac{c(a + b) - 2(a - b) \pm \sqrt{(c(a + b) - 2(a - b))^2 - 4c(cab - a^2 + b^2)}}{2c}, \tag{A.102}$$

where subscripts 1, 2 correspond to the $\pm$ signs in the right-hand side.

If we expand all the products and squares in the discriminator and collect the terms, the solution simplifies to the following:

$$x_{1,2} = \frac{c(a + b) - 2(a - b) \pm (a - b)\sqrt{c^2 + 4}}{2c}. \tag{A.103}$$

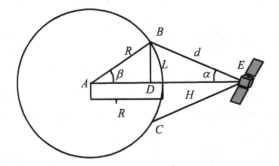

Figure A.6
Designing satellite coverage

A.24 Designing Satellite Coverage

Problem

Figure A.24 shows a cross section of this problem. A satellite flies at altitude H above Earth and transmits a signal downward. The antenna beam has a conical shape, and the axis of the cone is directed to Earth's center. The angle from the center to the edge of the beam is α. Earth's radius is R. On Earth's surface, the satellite beam covers a round spot, whose diameter spans the distance between points B and C. We must compute the length of the arc on Earth's surface from the center of the beam spot to its edge.

Solution

The angle from the beam center to its edge as seen from the Earth center is denoted by β. The distance from the satellite to the rim of the beam spot on Earth's surface is denoted by d. We formulate two equations for β and d using standard geometric formulas:

1. *The first equation for β and d.* From the law of sines for triangle ABE, we get

$$R \sin \beta = d \sin \alpha. \tag{A.104}$$

2. *The second equation for β and d.* The distance from the center of Earth to the satellite is the sum of the legs of two right triangles, ABD and DBE. These legs can be computed using the lengths of the hypotenuses and the definition of cosine:

$$AD = R \cos \beta,$$
$$DE = d \cos \alpha. \tag{A.105}$$

Separately, the sum $AD + DE$ is equal to $R + H$. Therefore,

$$R \cos \beta + d \cos \alpha = R + H. \tag{A.106}$$

Equations (A.104) and (A.106) serve as a basis to compute β. We substitute d from equation (A.104) in equation (A.106) to get

$$R\cos\beta + R\sin\beta\frac{\cos\alpha}{\sin\alpha} = R + H. \tag{A.107}$$

This yields

$$\sin\alpha\cos\beta + \sin\beta\cos\alpha = \frac{R+H}{R}\sin\alpha. \tag{A.108}$$

In the left-hand side we recognize the trigonometric identity for the sine of a sum of two angles.[5] The last equation takes the following form:

$$\sin(\beta + \alpha) = \frac{R+H}{R}\sin\alpha. \tag{A.109}$$

After solving for β, we get

$$\beta = \sin^{-1}\left(\frac{R+H}{R}\sin\alpha\right) - \alpha. \tag{A.110}$$

This angle is key to determining the size of the beam on Earth's surface. The length of the arc from the center to the edge of the beam on Earth is given by

$$L = \beta R = R\left(\sin^{-1}\left(\frac{R+H}{R}\sin\alpha\right) - \alpha\right). \tag{A.111}$$

A.25   Detecting a Vessel by Two Radars

Problem

A radar measures a range (distance) to the object it is tracking. Assume that there are two radars that detect a sea vessel at ranges R_1 and R_2. If the respective coordinates of the radars are $x_1 = 0; y_1 = 0$ and $x_2 = D; y_2 = 0$, where is the detected vessel located?

Solution

The geometry of the problem is shown in figure A.7. We denote the coordinates of the sea vessel by (x, y). From the Pythagorean theorem we get

$$\begin{aligned} R_1^2 &= x^2 + y^2, \\ R_2^2 &= (D-x)^2 + y^2. \end{aligned} \tag{A.112}$$

We substitute y^2 from the first equation in the second one:

$$R_2^2 = (D-x)^2 + R_1^2 - x^2. \tag{A.113}$$

Expanding the square produces a linear equation for x:

$$R_2^2 = -2xD + D^2 + R_1^2. \tag{A.114}$$

5. The sine of a sum of two angles: $\sin\alpha\cos\beta + \sin\beta\cos\alpha = \sin(\alpha + \beta)$.

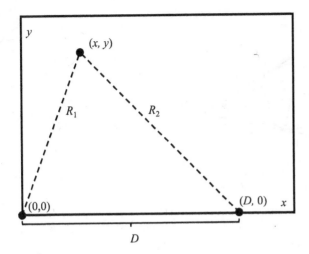

Figure A.7
Detecting a vessel by two radars

Its solution is as follows:

$$x = \frac{D^2 + R_1^2 - R_2^2}{2D}.$$

(A.115)

Plugging this into the first equation of the original system yields the solution for y:

$$y = \pm \sqrt{R_1^2 - x^2}.$$

(A.116)

A.26 Two Circles Inscribed in an Angle

Problem

Two circles are inscribed in an angle in such a way that they touch each other (figure A.8). What is the ratio of their radii as a function of the angle?

Solution

We denote the distance from the vertex of the angle to the center of the smaller circle as d and to the larger circle as D. The respective radii of the circles are r and R. The half-angle measure is α. From figure A.8 we see that

$$D = d + r + R,$$
$$\frac{R}{D} = \sin \alpha,$$
$$\frac{r}{d} = \sin \alpha.$$

(A.117)

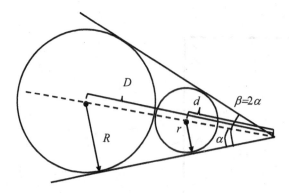

Figure A.8
Two circles inscribed in an angle

We divide the first equation by d to get

$$\frac{D}{d} = 1 + \frac{r}{d} + \frac{R}{d}.$$

(A.118)

The second term r/d in the right-hand side is equal to $\sin \alpha$. The last term R/d is expressed as follows:

$$\frac{R}{d} = \frac{R}{D} \cdot \frac{D}{d} = \frac{D}{d} \sin \alpha.$$

(A.119)

Now equation (A.118) is in the form:

$$\frac{D}{d} = 1 + \sin \alpha + \frac{D}{d} \sin \alpha.$$

(A.120)

We solve it for D/d:

$$\frac{D}{d} = \frac{1 + \sin \alpha}{1 - \sin \alpha}.$$

(A.121)

This ratio is also equal to the desired ratio of the radii of the circles:

$$\frac{R}{r} = \frac{1 + \sin \alpha}{1 - \sin \alpha}.$$

(A.122)

To express the result in terms of the full angle, we recall that α is the measure of a half-angle and $\beta = 2\alpha$. Then

$$\frac{R}{r} = \frac{1 + \sin \frac{\beta}{2}}{1 - \sin \frac{\beta}{2}}.$$

(A.123)

A.27 A Circle Inscribed in a Right Triangle

Problem

A circle is inscribed in a right triangle (see figure A.9). One of the legs has length a, and the measure of the adjacent angle is α. What is the radius of the circle?

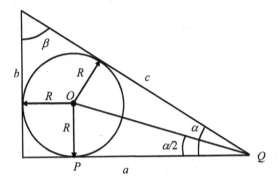

Figure A.9
A circle inscribed in a right triangle

Solution

We denote the circle radius as R. From triangle OPQ in figure A.9 we can see that

$$(a - R)\tan\frac{\alpha}{2} = R. \tag{A.124}$$

Solving this equation for R produces the following:

$$R = a\frac{\tan\frac{\alpha}{2}}{1 + \tan\frac{\alpha}{2}}. \tag{A.125}$$

Next, we use a trigonometric identity[6] for the tangent of a half-angle. This yields a simplified result:

$$R = \frac{a\sin\alpha}{1 + \sin\alpha + \cos\alpha}. \tag{A.126}$$

A.28 A Rectangle Inscribed in a Right Triangle

Problem

A rectangle is inscribed in a right triangle (see figure A.10). One of the legs has length a. The side of the rectangle that is aligned with that leg has length d. At what value of d, if any, is the area of the rectangle maximized?

Solution

We denote the measure of the adjacent angle by α. The area of the rectangle is the product of the lengths of its sides. One of the sides is given by d; the length of the second one is found from triangle PQS:

$$q = (a - d)\tan\alpha. \tag{A.127}$$

---

6. The tangent of a half-angle: $\tan(\alpha/2) = \sin\alpha/(1 + \cos\alpha)$.

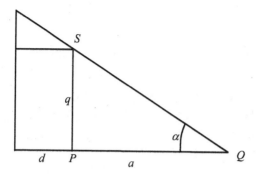

Figure A.10
A rectangle inscribed in a right triangle

Then area A of the rectangle is given by

$$A = d(a - d)\tan\alpha. \tag{A.128}$$

To find the value of d that maximizes area A, we complete the square in the right-hand side of the last equation:[7]

$$
\begin{aligned}
A &= d(a - d)\tan\alpha \\
&= -(d^2 - ad)\tan\alpha \\
&= -\left(d^2 - ad + \frac{a^2}{4} - \frac{a^2}{4}\right)\tan\alpha \\
&= -\left(\left(d - \frac{a}{2}\right)^2 - \frac{a^2}{4}\right)\tan\alpha.
\end{aligned}
\tag{A.129}
$$

The expression $(d - a/2)^2$ in the right-hand side is nonnegative and takes a minimum when $d = a/2$. This corresponds to the *maximum* for A (note the minus sign in the front of the right-hand side, which turns a minimum into a maximum). Therefore, the maximum area is achieved when the side d of the rectangle spans a half of the leg a.

A.29   The Cubic Formula

Problem

Solve the cubic equation for x:

$$ax^3 + bx^2 + cx + d = 0. \tag{A.130}$$

---

7. Calculus offers a better way to find extrema of functions, but it is beyond the scope of this book.

Solution

The solution is presented here without a derivation. We denote

$$\Delta_0 = b^2 - 3ac,$$
$$\Delta_1 = 2b^3 - 9abc + 27a^2d,$$
$$C = \sqrt[3]{\frac{\Delta_1 \pm \sqrt{\Delta_1^2 - 4\Delta_0^3}}{2}}. \tag{A.131}$$

Then one of the roots is given by

$$x_1 = -\frac{1}{3a}\left(b + C + \frac{\Delta_0}{C}\right), \tag{A.132}$$

where the plus or minus sign in this solution can be chosen arbitrarily, unless $C = 0$ for one of the signs, in which case we must choose the sign that yields a nonzero value of C.

For the so-called *depressed cubic equation*

$$x^3 + px + q = 0, \tag{A.133}$$

all three roots can be expressed through the trigonometric solution:

$$x_k = 2\sqrt{-\frac{p}{3}} \cos\left(\frac{1}{3}\cos^{-1}\left(\frac{3q}{2p}\sqrt{-\frac{3}{p}}\right) - \frac{2\pi k}{3}\right), \tag{A.134}$$

where $k = 0, 1, 2$.

A.30   A Spherical Cap

A spherical cap is the part of a sphere that lies above a plane that crosses this sphere (figure A.11).

Problem

Find the volume and the surface area of a spherical cap.

Solution

The solution is presented here without a derivation. The volume V and the surface area S (net of the base) are given by

$$V = \frac{1}{3}\pi h^2(3R - h),$$
$$S = 2\pi R h. \tag{A.135}$$

A.31   Mortgage Payments

Problem

This problem deals with computing payments for a so-called fixed-rate mortgage. This type of loan works in the following way:

1. A person or persons borrow some amount from a bank, often to buy a house. We denote the borrowed amount as D (for example, measured in dollars).

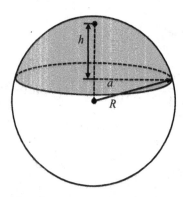

Figure A.11
A spherical cap

2. The bank requires that this loan is paid back with interest. We denote the interest rate as r. The mortgage is repaid over some fixed time period, which we denote as T.

3. The interest rate does not change over the duration of the mortgage. This is why this particular financial instrument is called the fixed-rate mortgage.

4. All payments are of equal amounts.[8]

5. Each payment consists of two parts:

   (a) One part goes to repaying the balance of the loan (called the *principal*). As time progresses, the loan balance is gradually decreased because of these payments; after time T it should hit zero.

   (b) The second part goes to paying the interest. For each payment, the interest is computed on the current remaining balance. Since the balance decreases over time, the interest portion in each payment gradually decreases as well.

   The mortgage is structured in such a way that all payments are equal, so that the sum of interest and principal payments remains constant. Since the interest portion decreases over time, the portion that goes to the principal must increase. In practice, the initial payments mostly go to paying interest, but over the years, they shift more and more to paying off the balance.

   This creates a tricky mathematical problem: what is the amount of the periodic payments that will pay off the original loan D over time T with the interest rate r?

Solution

The solution is presented here without a derivation. We provide an approximate formula for the payment rate:

---

[8]. In practice, these payments may have insurance and some taxes rolled into them, which can vary from year to year. For simplicity, we do not consider these extraneous factors. We concentrate on paying off the loan and on paying interest.

$$p \approx \frac{rDe^{rT}}{e^{rT} - 1},$$ (A.136)

where p is measured in dollars per year.

The difference between this approximate formula and the exact one is that in reality a mortgage is paid at discrete time intervals, usually monthly, but the approximate solution assumes that it is paid continuously, as if there were a steady drain from the borrower's bank account. For example, a borrower must pay \$2,000 on the first of each month for her mortgage for the total of \$24,000 per year, but the approximate formula assumes that the latter amount is gradually withdrawn over the course of the year. The approximate equation (A.136) is simpler than the exact formula, but still provides good accuracy.

A.32   The Kalman Filter

Problem

This problem is the most basic formulation of the famous Kalman filter algorithm, which has been used widely in science and engineering, including getting astronauts to the Moon.

The problem deals with processing measurements when we must account for a possibility of having an unknown error in each measurement.[9] In practice, we do not know the exact value of the measurement error in each case, but we usually know how big it *might* be for a particular measurement type or a particular instrument. For example, if you measure the length of a table using a measuring tape, you can never align the end of the tape with the end of the table perfectly; this introduces a small error in the result, which may be on the order of 1/16th of an inch. In practice, this error may be positive or negative, its value is unknown for each individual measurement, but we do not expect its magnitude to be three or more times greater than 1/16th of an inch.

In statistics, the typical magnitude of error is characterized by the so-called *error variance* σ^2. For the purposes of this section, you do not have to understand all the details about error variances; it is sufficient to accept that lower-quality measurements have larger variances and higher-quality measurements have smaller variances.[10] We also note that the variance is always nonnegative (the square in the notation σ^2 makes this explicit). If a quantity is measured in meters (m), its variance has units of m^2, and so on.

Suppose that you make two measurements of the same quantity using different measuring instruments. Both instruments are not perfect, and each measurement may contain some small measurement error. The first instrument produced a value X_1, and this measurement has a variance σ_1^2. The second measurement produced a value X_2, with a variance σ_2^2. In the above example, we may have

---

9. A measurement error is a normal part of any measurement process and does not imply a human error. That is why, when talking about an error in a measurement, we do not see the Pinocchio icon here, as we do elsewhere in this book for erroneous results.

10. If you do want to see a definition, variance σ^2 of a random variable X is defined as the expected value of the squared difference of that variable and its expected value μ:

$$\mu = E[X],$$
$$\sigma^2 = E[(X - \mu)^2],$$

where E[] denotes the expected value of a random variable in the brackets. The terms "expected value" and "random variable" are defined in section 7.1.

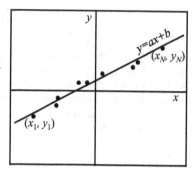

Figure A.12
Linear regression

measured the length of a table with a regular measurement tape (a less accurate instrument) and a laser interferometer (a more precise instrument). What is the best estimate for the measured quantity (for example, the length of the table), as derived from the two measurements available?

Solution

The best estimate (technically known as the *unbiased, minimum variance estimate*) for the quantity that is being measured is given by

$$X = \frac{X_1 \sigma_2^2 + X_2 \sigma_1^2}{\sigma_2^2 + \sigma_1^2}.$$

(A.137)

Since each of the contributing measurements may have an error in it, the value of X is generally not error-free either. Its accuracy is characterized by its own variance:

$$\sigma^2 = \frac{\sigma_2^2 \sigma_1^2}{\sigma_2^2 + \sigma_1^2}.$$

(A.138)

The derivation of this result requires knowledge of statistics, which is beyond the scope of this book.

A.33   Linear Regression

Problem

Linear regression is an algorithm that is often used to interpret a set of numerical data. In its simplest version, we deal with N measurements for a dependent variable y_i that were collected for corresponding values of an independent variable x_i. We assume that these two variables are linked by a linear relationship and that in addition there may be some measurement errors or random variations in the values of y:

$$y = ax + b + R,$$

(A.139)

where R is a random variable. The linear regression algorithm estimates the best fit for model parameters a and b from the available data.

For example, the management of a company tries to analyze the sales data. They plot sales numbers versus time and see a generally linear trend, with some deviations from it (see figure A.12). They assume a linear growth model for the sales, apply linear regression to estimate a and b, and then use that model to predict the sales figure for the next month.

Solution

The equations for the best fit for model parameters a and b are given here without a derivation:

$$a = \frac{N \sum_{i=1}^{N} x_i y_i - \sum_{i=1}^{N} x_i \cdot \sum_{i=1}^{N} y_i}{N \sum_{i=1}^{N} x_i^2 - \left(\sum_{i=1}^{N} x_i\right)^2},$$

$$b = \frac{\sum_{i=1}^{N} y_i \cdot \sum_{i=1}^{N} x_i^2 - \sum_{i=1}^{N} x_i \cdot \sum_{i=1}^{N} x_i y_i}{N \sum_{i=1}^{N} x_i^2 - \left(\sum_{i=1}^{N} x_i\right)^2},$$

(A.140)

where x_i, y_i are the available data for variables x and y.

Further Reading

To explore the subject of this book further, you may want to check out the following publications:

1. More about dimensional analysis can be found in Bridgman (1922) and Santiago (2019).

2. A great book by Barenblatt (2009) expands on dimensional analysis and goes into a deep dive on scaling.

3. Small parameter expansion and successive approximations are presented in Bender and Orszag (1999).

4. Basic practical usage of limiting cases and symmetry for calculus problems is found in Cipra (2000).

5. You can learn many tricks of the trade, including making order of magnitude estimates and limiting cases, from Mahajan (2010) and Mahajan (2014).

6. Bernstein and Friedman (2009) has a wealth of information on dimensional analysis, limiting cases, symmetry, estimation, and the perturbation theory.

7. The classic book on symmetry by Weyl (2016) is a gem.

Bibliography

Barenblatt, G. 2009. *Scaling, self-similarity, and intermediate asymptotics*. Cambridge: Cambridge University Press.

Bender, Carl M., and Steven A. Orszag. 1999. *Advanced mathematical methods for scientists and engineers: Asymptotic methods and perturbation theory*. New York: Springer.

Bernstein, M. A., and W. A. Friedman. 2009. *Thinking about equations: A practical guide for developing mathematical intuition in the physical sciences and engineering*. Hoboken, NJ: Wiley.

Bridgman, P. W. 1922. *Dimensional analysis*. New Haven, CT: Yale University Press.

Cipra, B. 2000. *Misteaks ... and how to find them before the teacher does ...: A calculus supplement*. Natick, MA: A. K. Peters.

Mahajan, S. 2010. *Street-fighting mathematics: The art of educated guessing and opportunistic problem solving*. Cambridge, MA: MIT Press.

Mahajan, S. 2014. *The art of insight in science and engineering: Mastering complexity*. Cambridge, MA: MIT Press.

Misra, P., and P. Enge. 2010. *Global positioning system: Signals, measurements, and performance*. Lincoln, MA: Ganga-Jamuna Press.

Santiago, J. G. 2019. *A first course in dimensional analysis: Simplifying complex phenomena using physical insight*. Cambridge, MA: MIT Press.

Sornette, D. 2006. *Critical phenomena in natural sciences: Chaos, fractals, self-organization and disorder: Concepts and tools*. Berlin: Springer.

Weyl, H. 2016. *Symmetry*. Princeton, NJ: Princeton University Press; reprint edition.

Index

Bold page numbers refer to a definition, introduction, or explanation of a term.

Index of Problems

Bold page numbers refer to the main formulation of a problem.